HETEROCYCLIC CH

HETEROCYCLIC CHEMISTRY

Second Edition

J. A. Joule
G. F. Smith
Department of Chemistry, University of Manchester

VAN NOSTRAND REINHOLD COMPANY
LONDON
NEW YORK CINCINNATI TORONTO MELBOURNE

First published 1972
Second edition 1978
Reprinted 1979

**Published by Van Nostrand Reinhold Company Ltd.,
Molly Millars Lane, Wokingham, Berkshire, England**

*Published in 1978 by Van Nostrand Reinhold Company
A Division of Litton Educational Publishing, Inc.,
450 West 33rd Street, New York, N.Y. 10001, U.S.A.*

*Van Nostrand Reinhold Limited
1410 Birchmount Road, Scarborough, Ontario, M1P 2E7,
Canada*

*Van Nostrand Reinhold Australia Pty. Limited
17 Queen Street, Mitcham, Victoria 3132, Australia*

Library of Congress Cataloging in Publication Data
Joule, John Arthur.
  Heterocyclic chemistry.
  Bibliography: p.
  Includes index.
  1. Heterocyclic compounds. I. Smith, George
Fouet, joint author. II. Title.
QD400.J59 1977 547'.59 75-186763
ISBN 0-442-30211-8
ISBN 0-442-30212-6 pbk.

Printed in Great Britain
by W & J Mackay Limited, Chatham

Introduction

This textbook is designed for the Honours student and for the advanced graduate who wishes to brush up the fundamentals of heterocyclic chemistry and obtain a better grasp of the reactivity of heteroaromatic systems, especially in the broader context of homoaromatic and general aliphatic chemistry.

The emphasis in this treatment of the subject is very much on heteroaromatic systems, for it is here that most of the lessons which are specifically heterocyclic can be learned and much that seems new can, on closer scrutiny, be seen to be already familiar and related to the reactivity of simpler systems.

In order better to drive the basic lessons home, only the simple well-known heteroaromatics are treated in depth: some of the more peripheral ones are described more or less briefly, with an emphasis on similarities and differences with the more fundamental systems.

In the one chapter on non-aromatic heterocycles we have in the main brought out those aspects in which they differ from their acyclic counterparts in aliphatic chemistry.

The book begins with a chapter on the structures of the heteroaromatics: the static picture of the molecule is given in some detail in the familiar bond-resonance terms, which in the opinion of many university teachers still is the best approach for most purposes. This is followed by a brief comment on the place of quantum mechanical molecular orbital theory in the teaching of heterocyclic chemistry.

Then there is another general chapter which explains the general principles and the main types of reaction used in the synthesis of heteroaromatic systems.

The student is strongly advised to read these initial chapters carefully, for much that is said there is subsequently assumed. In his subsequent studies, the reader will find it helpful to refer back to these sections.

The bulk of the subject matter deals with the chemistry of pyridines, quinolines, isoquinolines, the three diazines, pyrylium salts, the pyrones, benzopyrylium salts, the benzopyrones, pyrroles, furans, thiophens, indoles, 1,3-azoles, and the purines.

The systems treated more briefly are quinolizinium and thiopyrylium salts, isoindole, indolizine, benzofuran, benzothiophen, the 1,2-azoles, and the non-aromatic heterocycles.

To have included more systems would, we feel, have detracted from the value of the treatment we have worked out. A reading list at the end of each main section and at the end of the book lead to aspects which the reader may wish to study in greater depth.

In this main section of the book, the chemistry of each heterocycle is first discussed in one chapter in general terms, that is, broad mechanistic aspects are presented and the relationship of the reactivity of the system to other appropriate systems brought out; then, in a following chapter, the chemistry is discussed systematically in detail. In these detailed chapters the same sequence of reaction types is largely adhered to, so that an introduction giving general information including the main biologically and chemotherapeutically significant molecules is always followed by reactions with electrophilic reagents in a standard order: protonation, nitration, sulphonation, etc., then by reactions with oxidizing agents, with nucleophilic agents, with free radicals, with reducing agents, and with dienophiles; then come reactions of metalloderivatives, of alkyl, carbonyl, halo, oxy, and amino derivatives. Where a particular aspect has nothing of special interest to offer, it is generally left out. The chapter then ends with a discussion of synthesis: here the most important general methods for the system are examined and a number of literature syntheses are given, without detailed comment, which illustrate both ring synthesis and substituent manipulation.

It is hoped that this two-stage treatment will make for better understanding and easier assimilation of information. Repetition in this approach is inevitable, but we believe that repetition of various important facts and generalizations in different contexts is didactically valuable.

Introduction to the Second Edition

The past five years have seen the publication of quite a number of important papers which have contributed to our better understanding of heterocyclic chemistry, and in the light of this work the text has been brought up to date. Owing to the rapid growth of research in organic photochemistry, a chapter dealing with the photochemistry of heteroaromatic systems has been added.

Contents

Glossary of Abbreviations and Symbols

Chemical symbols are used in the text, as well as more usually on reaction arrows, since we feel that this does not detract from clarity while economizing on space.

ny = no yield given in original account.

hy = original account describes yield as 'high', 'good', or 'quantitative'.

ly = original account describes yield as 'low'.

RT = experiment carried out at room temperature.

Δ = experiment carried out at reflux.

atm = atmospheres.

aq = aqueous solution.

(liq) = in liquid phase.

c = concentrated.

f = fuming.

Me = methyl.

Et = ethyl.

Pr^n = normal propyl.

Pr^i = isopropyl.

Bu^n = normal butyl.

Bu^t = tertiary butyl.

Ph = phenyl.

Ac = acetyl (CH_3CO-).

THF = tetrahydrofuran.

DMF = dimethylformamide.

DMSO = dimethylsulphoxide.

LAH = lithium aluminium hydride.

NBS = *N*-bromosuccinimide

AIBN = azobisisobutyronitrile

HMPA or HMPT = hexamethyl phosphoric triamide

LDA = lithium di-isopropylamide

LTMP = lithium 2,2,6,6-tetramethylpiperidide

1

Structure and Main Physical Properties of the Aromatic Heterocyclic Systems

This chapter presents in simple terms the valence-bond view of the main heteroaromatic systems, and also briefly assesses the current value of molecular orbital calculations in relation to elementary teaching of heterocyclic chemistry.

Brief descriptions of benzene, naphthalene, cyclopentadienyl anion and cycloheptatrienyl cation are included in order to provide a background and to emphasize the close relationship of carbocyclic and heterocyclic aromatic structures.

CARBOCYCLIC AROMATIC SYSTEMS

The concept of aromaticity as represented by benzene is a familiar and relatively simple one. We know well the difference in reactivity between benzene on the one hand and olefins like ethylene or, say, cyclohexadiene on the other: that is, that the olefins react rapidly *by addition* with electrophiles such as bromine, whereas benzene reacts only under much more forcing conditions and then nearly always *by substitution*. This difference is due to the cyclic arrangement of the six π-electrons in benzene: this forms a conjugated molecular orbital system which is thermodynamically much more stable than a corresponding non-cyclically conjugated system. This extra stabilization generally results in a diminished tendency to react by addition and a greater tendency to react by substitution, with survival of the original cyclic conjugated system of electrons in the product, and is characteristic of aromatic compounds.

A general rule proposed by Hückel in 1931 states that aromaticity is observed in cyclically conjugated systems of $4n + 2$ electrons, that is, with 2, or 6, or 10, or 14, etc., electrons.

The more recent and more fundamental molecular-orbital description of aromatic systems, which, most importantly, allows the essential aspects of their electronic absorption characteristics to be rationalized, is of course fully available. This aspect of theoretical chemistry is one of the most important present-day developments in chemical thinking. We believe, however, that it does not yet play an important part in the teaching of elementary aromatic chemistry: its fullest impact is yet to be felt in the field of reactions which proceed by so-called ionic mechanisms in solution (see also p. 22).

In this treatment of heterocyclic chemistry, therefore, we shall use the simpler and more pictorial valence-bond resonance description of structure and reactivity. Even though this treatment is far from rigorous, generations of students have found it a valuable aid to the understanding and learning of elementary organic chemistry, which at a much more advanced level gives way naturally to the much more complex quantum mechanical approach.

A brief valence-bond resonance description of benzene and of naphthalene follows in order to pave the way for a similar description of the heteroaromatic systems.

Valence-bond View of the Structure of Benzene and Naphthalene

In benzene, the geometry of the ring, with angles of 120°, precisely fits the geometry of a planar trigonally hybridized carbon atom, and allows the setting-up of a σ-skeleton of six sp^2 carbon atoms in a strainless planar ring: each carbon then has one extra electron which occupies an atomic orbital orthogonal to the plane of the ring. These electrons may be combined in spin-coupled pairs to produce a 'Kekule structure', 1 (1A).

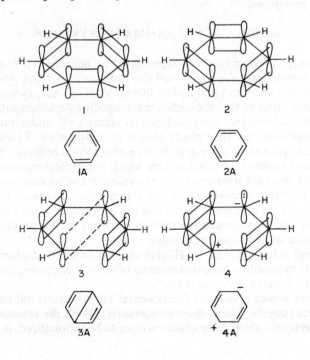

However, as has been known for over 100 years, there are two ways of combining electrons in pairs in this way, the second being 2 (2A): these two structures are entirely equivalent, and they interact, by what was known as bond resonance and more recently has been named exchange degeneracy, to lead to complete equivalence of all the C—C bonds. In more complete treatments of the structure, other modes of coupling the electrons are incorporated into the calculations, such as the six in which two electrons pair across the ring as in 3 (3A), or the six in which polarization occurs as in 4 (4A): but these are high-energy modes of coupling which make only relatively minor contributions to the overall picture. Canonical structure 3A superficially resembles that of bicyclo[2.2.0]hexa-2,5-diene (Dewar benzene), a well known and relatively stable hydrocarbon: the big difference of course lies in the normal bond-length at the ring junction of Dewar benzene, and its non-planar geometry (see Dewar pyridine, p. 352).

Structures 1–4 (1A–4A) are canonical forms: they have no physical existence as such, that is, benzene is never at any time like any of them but has a hybrid structure which is intermediate between them all.

Benzene thus emerges as a regular hexagon with a measured bond-length of 1·39 Å, which is intermediate between the 1·34 Å of the alkene double bond and the 1·46 Å of a single bond between sp^2 hybridized carbons in polyenes.

benzene

cyclopentadiene   hypothetical
                                   cyclohexatriene

When naphthalene is treated likewise, one finds that simple pairing of the ten π-electrons gives three canonical structures, 5, 6, and 7, which correspond to structures 1 and 2 of benzene; canonical structures 8 and 9 are of much higher energy and contribute little to the ground state structure of naphthalene. Just by taking the geometrical average of 5, 6, and

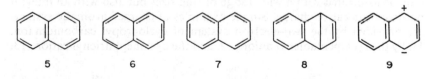

5   6   7   8   9

7, the C 1—C 2 bond comes to 1⅔ of a single bond; C 2—C 3, C 1—C 9, and C 9—C 10 each comes to 1⅓ of a single bond. Now, by simple proportion, a 1⅓ bond would be about 1·42 Å long, and a 1⅔ bond about 1·38 Å. Over the years, actual measurements of the bond-lengths of naphthalene have given a range of values, but if one takes one of the most recent sets shown in 10, published in 1966 and determined by electron-diffraction, the closeness of the measured values to those arrived at by taking the geometric mean of canonical structures 5, 6, and 7 is striking.

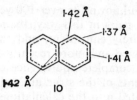

Aromatic Resonance Energy

The difference between the ground-state energy of benzene and that of a hypothetical non-aromatic 1,3,5-cyclohexatriene corresponds to the degree of stabilization conferred to benzene by special cyclic interaction of the six π-electrons. This difference is usually known as aromatic resonance energy.

A very important point to understand is that aromatic resonance energy is a very difficult property to measure and calculate. The difficulty lies in estimating the energy of the hypothetical non-aromatic structure. This estimate is least difficult to make in the case of benzene, becomes more difficult in the case of naphthalene, and still more difficult in the case of the simple heterocyclic compounds like pyridine, pyrrole, furan, etc. This is evident in the wide range of values published for each of the main heterocycles.

The resonance energy of systems like pyrylium, pyridinium, pyridones, and pyrones are the most difficult to assess.

At this stage perhaps it is best not to put a precise figure to aromatic resonance energies, but just to state in general terms that the resonance energy of pyridine is of the same order as that of benzene, that of thiophen is lower, perhaps closely followed by pyrrole, and that of furan is the lowest of all the simple uncharged heterocycles with one heteroatom.

Cyclopentadienyl Anion and Cycloheptatrienyl Cation

Aromatic planar cyclic systems of 2, 6, 10, etc., π-electrons may not only be associated with a wide range of ring sizes but also with an integral positive or negative charge. Simple examples of charged aromatic systems are provided by the two-electron system of cyclopropyl carbonium ion, 11, by the cyclopentadienyl anion, 12, and the cycloheptatrienyl cation, 13.

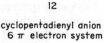

cyclopropyl carbonium ion   cyclopentadienyl anion   cycloheptatrienyl cation
2 π electron system   6 π electron system   6 π electron system

These are all highly reactive ions, but can be seen to be much less reactive than expected in the absence of stabilization by aromatic delocalization: thus cyclopentadiene, with a pK_a of 14–15, is very much more acidic and more easily deprotonated to the anion than expected of a simple diene. In the five equivalent main canonical forms of the anion, 14 (14A), 14', etc., the six electrons are paired as shown and the resulting resonance hybrid is a regular pentagon. It may be noted that in the σ-ring skeleton there must be appreciable strain due to compression of the sp^2 hybrid angle of 120° to 109° of the pentagon.

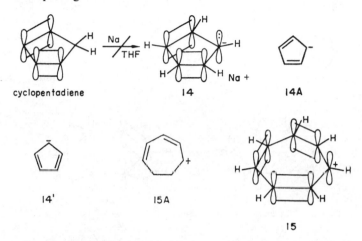

cyclopentadiene   14   14A

14'   15A   15

Tropylium bromide is an ionic, water-soluble compound: the cation has seven equivalent main canonical forms, the six electrons are paired as shown in 15 (15A), the seventh carbon having an empty orbital. The resonance hybrid then is a regular heptagon.

HETEROAROMATIC SYSTEMS

Pyridine and Related Systems

The structure of pyridine is completely analogous with that of benzene being related by replacement of ═CH— by ═N—. The differences are: (a) a departure from perfectly regular hexagonal geometry caused by the shorter C—N bond; (b) the replacement of a hydrogen in the plane of the ring by an unshared electron pair, likewise in the plane of the ring and in an sp^2 hybrid orbital, not at all involved in the aromatic π-electron system, and responsible for the basic properties of pyridines; (c) a strong permanent dipole, caused by the greater electronegativity of nitrogen compared with that of carbon.

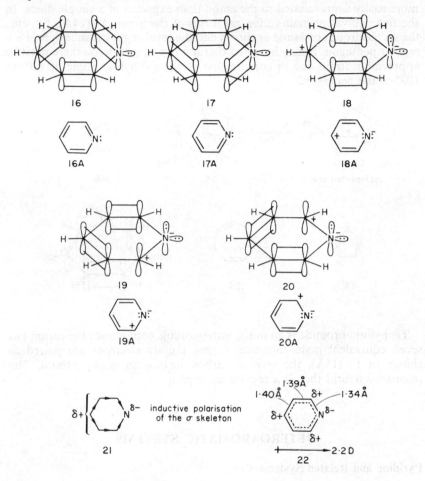

The more electronegative nitrogen causes both inductive polarization, mainly of the σ-bonds as shown in 21, and stabilizes those polarized canonical structures in which the nitrogen is negatively charged, 18, 19, and 20, which thus make a significant contribution to the hybrid structure. The main canonical forms are, as in benzene, the Kekule structures 16 and 17. Pyridine, then, is a molecule with essentially equivalent bonds in which both inductive and mesomeric effects work in the same direction and result in a dipole of 2.21 D, the negative end of which is on the nitrogen, the positive fractional charges being located mainly on C 2, C 4 and C 6. The dipole moment of piperidine gives an idea of the value of the induced polarization of the σ-skeleton of pyridine.

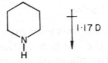

The structures of the *diazines* are analogous, thus pyrimidine can be represented by the following canonical structures:

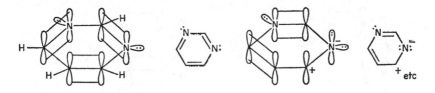

The structures of *quinoline* and *isoquinoline* bear the same relationship to pyridine as naphthalene does to benzene.

Pyridinium and Related Cations

Electrophilic addition to the pyridine nitrogen generates pyridinium ions, 23, the simplest being 1H-pyridinium (23, R=H) formed by inter-

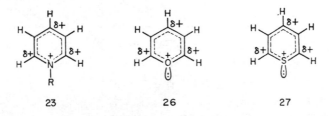

charge on hetero atom roughly 1–3δ

action with a protic acid. In fact 1H-pyridinium is isoelectronic with benzene, and the difference lies in the higher nuclear charge of N which makes the system positively charged. Thus pyridinium cations are still very much aromatic in character. The formal positive charge on the nitrogen, however, must interact quite strongly with the aromatic molecular orbital system and reduce its stabilizing effect: furthermore the positive charge is delocalized on to ring carbon both mesomerically (cf. pyridine 18, 19, 20) and inductively (cf. pyridine 21) and C 2, C 4 and C 6 carry fractional positive charges which are much higher than in pyridine.

Pyrylium and Thiopyrylium Cations

These, again, are closely analogous with pyridine and more so with pyridinium cation. In pyrylium, the oxygen carries an unshared electron pair in an sp^2 hybrid orbital in the plane of the ring, exactly as in pyridine. Being in an essentially tricovalent state, the oxygen is necessarily positively charged:

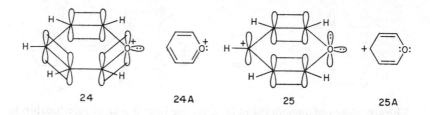

24   24A   25   25A

One of the Kekule canonical structures is represented in 24 (24A) and one of the three main canonical structures in which the positive charge is on carbon is represented in 25 (25A): the structure of the hybrid then may be given as 26, in which the formal positive charge is more strongly delocalized on to carbon than it is in pyridinium. In fact, being so much more electronegative than nitrogen, oxygen tolerates a formal positive charge much less than does nitrogen. Pyrylium is probably stabilized by aromatic delocalization to a lesser extent even than pyridinium. However, no experimental data at all are available bearing on this point.

Thiopyrylium, 27, is structurally very close to pyrylium.

Pyrrole

Pyrrole is isoelectronic with cyclopentadienyl anion, and is electrically neutral because of the higher nuclear charge of N. The other consequence of the presence of N in the pyrrole ring is loss of the radial symmetry of cyclopentadienyl anion, so that pyrrole cannot have five exactly equivalent canonical forms but instead has one in which there is no charge separation, 28 (28A), and two pairs of equivalent forms, in which there is charge separation, 29 (29A)–30 (30A) and 31 (31A)–32 (32A). It is fairly obvious that these three different structures will have different energies, and that the actual hybrid structure will be made up of different contributions from each in proportion to their relative stability: the order of importance is believed to be 28 > 29, 30 > 31, 32. Resonance, then, leads to the establishment of partial negative charges on the carbon atoms, and a partial

4n+2   6
2

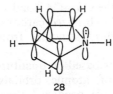

28

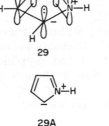

29

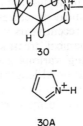

30

28A   29A   30A

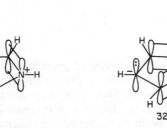

31   32

31A   32A

positive charge on nitrogen equal in value to the sum of the partial negative charges, as shown in 33. The inductive effect of nitrogen, however,

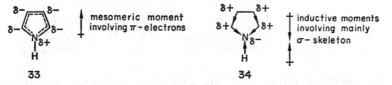

33   34

will cause a polarization towards nitrogen, mainly in the σ-bonded skeleton,
as shown in 34. The electron distribution in pyrrole, then, is a balance of
two opposing effects, of which the mesomeric effect probably is the stronger.

The actual measured dimensions of the C—C bonds of pyrrole show

35

the C 2—C 3 and C 4—C 5 bonds to be significantly longer at 1·37 Å than the double bond of cyclopentadiene at 1·34 Å, and the C3—C4 bond of pyrrole to be shorter than a single bond between sp^2 hybridized carbons: this indicates appreciable contributions to the hybrid structure from canonical structures 29–32.

In recent years very many sophisticated calculations based on quantum mechanical treatment of molecular orbitals and of atomic orbitals, making various assumptions and simplifications, have produced theoretical bond lengths very close to the observed ones.

It is very important to remember that the lone-pair of electrons of the N atom forms part of the aromatic π-electronic system.

Furan

Furan is clearly a close analogue of pyrrole, to which it is related by replacement of —NH— by —O—. The N-hydrogen of pyrrole, then, is replaced by an unshared electron pair in an sp^2 hybrid orbital, which, being in the plane of the ring, does not participate at all in the aromatic sextet of electrons and resembles the unshared electron pair of the pyridine nitrogen.

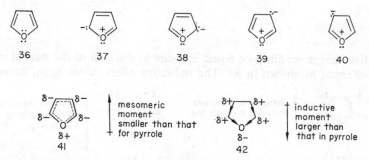

The canonical forms 36–40 are completely analogous with those of pyrrole (28A–32A). An important difference is the consequence of oxygen being more electronegative than nitrogen. This makes trivalent positively charged oxygen less stable than tetravalent positively charged nitrogen, and makes the canonical structures 37–40 less stable than the corresponding pyrrole structures (29A–32A): the contributions which structures 37–40 make to the resonance hybrids are then less important than in the pyrrole case, so that the hybrid is closer to canonical structure 36 than pyrrole is to structure 28A. This is reflected in the measured C—C bond lengths of furan which lie half-way between those of cyclopentadiene and those of pyrrole.

An additional consequence of the greater electronegativity of oxygen is the greater induced polarization, which is so strong as to lead to an overall dipole moment for furan with the negative end on oxygen (see p. 12).

All in all, mainly by reducing the extent of mesomerism, or resonance, the electronegativity of oxygen has the effect of making furan less aromatic in character than pyrrole.

Thiophen

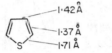

Thiophen is the sulphur analogue of furan, and its structure is very similar. Thiophen, however, has a higher degree of stabilization energy than furan, and is more aromatic in character as is borne out by its chemical reactions. There are several possible reasons for this: one is that sulphur has a larger bonding radius than oxygen which allows the C—C—C bond-angles to widen a little, thus allowing some gain in stability by

losing a little angle-compression strain; a second reason is that trivalent sulphur tolerates a positive charge better than oxygen, so that polarized canonical forms related to 37–40 contribute more to the resonance hybrid, with corresponding increase in resonance stabilization energy; a third reason could be that sulphur, being a second-row element, can use d-orbitals for bonding, which means that in bond-resonance terms, one can envisage contributions from canonical forms such as 43–45.

43   44   45

The question of how, and even if, d-orbitals are involved in quantum-mechanical molecular orbital terms is still under discussion.

The sulphur atom carries an electron pair in an sp^2 hybrid orbital in the plane of the ring (cf. furan).

The Dipole Moments of Pyrrole, Furan, and Thiophen

It is instructive to look at the dipole moments of these systems in greater detail, for most textbooks erroneously place the positive end of the dipoles of all three on the heteroatom, presumably because one instinctively feels that in these very reactive systems the overall excess negative charge should be on the highly nucleophilic carbons. As already indicated, however, in both furan and thiophen the negative end of the dipole is on the heteroatom.

In order to understand this situation, let us first consider the dipole moments of the saturated compounds pyrrolidine, tetrahydrofuran, and tetrahydrothiophen: all three have the negative end at the heteroatom. This polarization is the consequence of the inductive pull of the more electronegative heteroatom on the σ-bonding electrons.

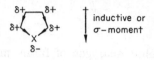

In pyrrole, furan, and thiophen this inductive or σ-moment is still operating towards the heteroatom, but superimposed on it is the mesomeric moment, mainly involving the π-electron system, operating in the opposite direction.

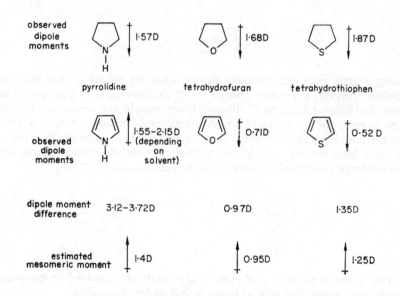

The observed dipole in furan and thiophen is largely the resultant of these two opposite polarizations: the inductive effect is the stronger, so that the heteroatom is still at the negative end of the observed dipole. The situation in pyrrole and pyrrolidine is complicated by the polarization of the N—H bond (and of the N-alkyl bond in N-alkylpyrroles): The very large difference of over 3 D between the dipole moments of pyrrole and pyrrolidine is made up of the mesomeric moment, which has been estimated to be about 1·4 D, and of other contributions, the main one of which must be the much stronger N—H ←+ dipole of the pyrrole system.

Thus it is important to realize fully that the excess negative charge placed on the carbons of furan and thiophen by molecular orbital calculations is not inconsistent with these carbons being at the positive end of a dipole, for the excess negative charge refers to the π-electrons, and not to the electronic system as a whole.

Indole

The structure of indole is a straightforward combination of benzene and pyrrole, the main canonical forms being 46–50. Less important canonical forms carry the negative charge on the various benzene positions. The

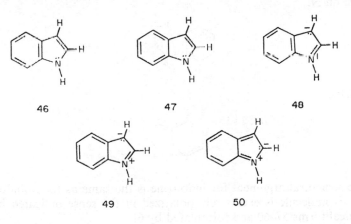

46   47   48

49   50

hybrid structure then carries a considerable fractional positive charge on the nitrogen.

Isoindole

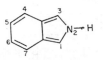

In the structure of isoindole, which is isoelectronic with indole, the polarized canonical forms 52–55 are probably close to the non-polarized canonical structure 51 in energy, for 51 is non-benzenoid, which offsets the advantage of not having charge separation, and 52 to 55 are benzenoid, which makes up for the disadvantage of charge separation.

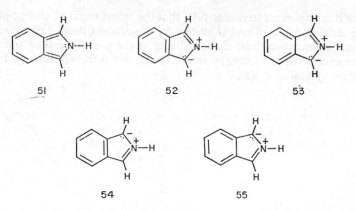

51   52   53

54   55

The hybrid structure then will carry very considerable fractional negative charges at C 1 and C 3, with a corresponding fractional positive charge on N.

Indolizine

The structural argument for indolizine is the same as for isoindole, so that this molecule is extensively polarized in the sense indicated by the canonical forms 57–60 and formulated by 61.

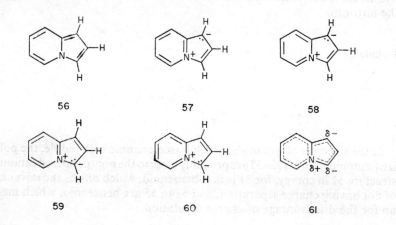

56   57   58

59   60   61

Imidazole

The structure of imidazole is very interesting, and combines the features both of pyridine and of pyrrole: it can be looked on as pyrrole in which a -β-CH— has been replaced by —N=, an azomethine nitrogen.

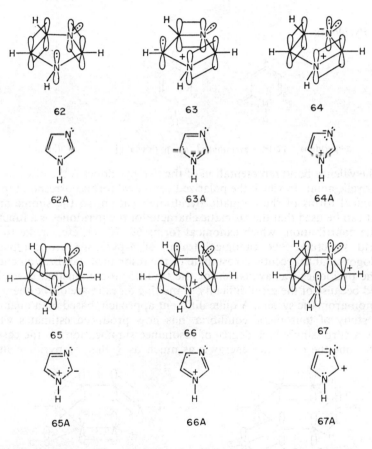

The five main ways in which the six electrons can pair are the same as in pyrrole and furan, with the difference that one of the canonical forms, 64, contributes much more to the final hybrid structure because a negative charge is much more stabilized on nitrogen than on carbon. A sixth canonical form, 67, probably also plays a significant role, the positive charge on C 2 being stabilized by the formal negative charge on N 3, much as in the pyridine canonical forms 19 and 20. The inductive pull of both nitrogens must greatly reduce electron-availability at the ring carbons: C 2 is the most strongly affected in this way, since it is also depleted by mesomerism.

Imidazole, then, has a very considerable fractional negative charge on N 3 and a partial positive charge on N 1.

The Pyridones

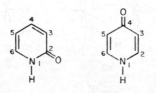

a-pyridone   [2(1H)-pyridinone]   γ-pyridone   [4(1H)-pyridinone]

The valence-bond representation of the two pyridones is the obvious one of a cyclic amide in which the polarized canonical forms correspond to the canonical forms of the aromatic pyridinium cation, so the simple argument can be used that the aromatic character of the pyridones is a function of the contributions which canonical forms 69, 70, 71, etc., make to the hybrid structure. The canonical forms of 4-pyridone are completely analogous. Until recently no estimate of the resonance stabilization energy of the pyridones was available, for the traditional combustion procedure failed because of the great difficulty of making an estimate of the energy of the non-aromatic system. A quite different approach, based on a quantitative study of tautomeric equilibria, has now produced estimates which show a surprisingly high degree of resonance stabilization: in the case of α-pyridone the resonance energy is as much as $\frac{3}{4}$ that of pyridine itself.

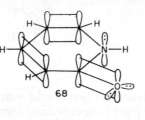

68   69

68A   69A

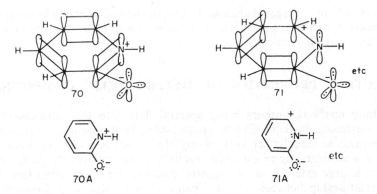

In this connection it is surprising to note that the C=O stretching frequency of 2-pyridone is about 1650 cm⁻¹, and that of N-methyl-2-pyridone is even higher, at 1690 cm⁻¹, these values falling well into the range for simple amides: a high contribution from polarized forms 70 and 71 would have led one to expect very much lower frequencies. There are many unanswered questions in connection with the structure of the pyridones. Indeed, the same is the case with the detailed analysis of the structure of the simple amide group.

The Pyrones

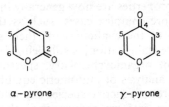

The structures of the pyrones are formally closely analogous with the structures of the pyridones as expressed in 68 (68A) to 71 (71A), with the very important difference that the indications from spectral properties and chemical reactivity show that they are probably not stabilized to any great extent by aromatic resonance, 2,6-dimethyl-4-pyrone, 72, for example, has an ultraviolet absorption which is similar to that of the dihydro-

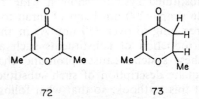

derivatives, 73 (72, λmax 247 nm, ε 14000; 73, λmax 263 nm, ε 10500). The carbonyl stretching frequencies of both α- and γ-pyrones are close to those of cyclically unconjugated analogues, but, in view of what has been said in connection with the pyridones, this cannot be used as an argument in favour of low aromaticity.

SPECTRAL PROPERTIES OF HETEROCYCLIC COMPOUNDS

Many textbook authors bring spectral data into their discussion of heteroaromatic systems. Clearly, spectroscopy is of enormous importance in research and analytical work of any kind, and the very rapid advances in our understanding of chemical reactivity of the last twenty years or so are to a large extent based on spectral analysis: we nevertheless feel that the relationship between, say, a τ value, or a λ value, and chemical reactivity is not sufficiently direct to be of value to an elementary student. The correct appreciation of such data involves quite specialized knowledge which is not within the teaching scope of this book, and is generally only acquired when pursuing research in the heterocyclic field: a brief description of the UV and NMR spectra of the main heteroaromatic systems is given in the following section, and those readers wishing to pursue this aspect of the subject further are recommended to read the relevant reviews quoted at the end of this textbook.

Suffice it to say here that the Ultra-violet, Infra-red and Nuclear Magnetic Resonance spectra of the simple heteroaromatic systems are in full accord with their aromatic character and that, indeed, theoretical calculations of electronic absorption properties are now generally in good agreement with observation. In the more complex cases, such as the pyridones and the pyrones, the question of the evaluation of their spectral properties in terms of, say, resonance stabilization, is still being debated.

As in other areas of organic chemistry, spectral data have been of particular value in the analysis of tautomeric equilibria (e.g., those of the pyridones), the structures of reactive species and intermediates (e.g., those of 3H-indolium cations), rate measurements (such as those of the nitration of quinoline), pK_a measurements, and in the exploration of new reactions. The analytical methods involved in such studies are quite general, and do not need to be described in this textbook.

Ultra-violet Spectra of Heteroaromatic Compounds

The simple unsubstituted systems show a wide range of electronic absorption, from the simple 200 nm band of furan to the complex spectrum of indolizine reaching to over 350 nm. As in the case of benzene compounds, the introduction of substituents such as —OH, —CHO, —NO$_2$, etc., into these systems causes profound changes in electronic absorption: an adequate description of such substituent effects is really beyond the scope of this textbook, so that what follows limits itself to a brief discussion of the spectra of the simple unsubstituted heteroaromatic systems.

The spectra of the monocyclic azines show two bands, each with fine structure: one occurs in the relatively narrow range of 240–260 nm and corresponds to the $\pi \to \pi^*$ transitions, which are analogous with the $\pi \to \pi^*$ transitions occurring in the same region in benzene; the other band occurs at longer wavelengths, from 270 nm in pyridine to 340 nm in pyridazine, and corresponds to interaction of the nitrogen lone pair of electrons with aromatic π-electrons, the $n \to \pi^*$ transitions, which of course cannot occur in benzene. The absorption due to $n \to \pi^*$ transitions is very solvent-dependent, as is shown in the table in the case of pyrimidine; in the case of pyridine, this absorption is only observed in hexane solution, for in alcohol solution the shift to shorter wavelengths results in masking by the main $\pi \to \pi^*$ band. Protonation of the ring nitrogen quenches the $n \to \pi^*$ band; protonation also has the effect of considerably increasing the intensity of the $\pi \to \pi^*$ band without changing its position significantly, which has considerable diagnostic value.

Table I   Monocyclic azines
(fine structure not given)

	$n \to \pi^*$		$\pi \to \pi^*$	
	λ_{max}(nm)	ϵ	λ_{max}(nm)	ϵ
pyridine (hexane)	270	450	195, 251	7500, 2000
pyridine (ethanol)			257	2750
pyridinium (ethanol)			256	5300
pyrazine (hexane)	328	1040	260	5600
pyridazine (hexane)	340	315	246	1300
pyrimidine (hexane)	298	326	243	2030
pyrimidine (H_2O)	271	420	243	3210
pyrimidinium (aqH_3O^+)			242	5500
pyrylium (70% $HClO_4$)			220, 269	1400, 8500
benzene (hexane)			204, 254	7400, 200

The bicyclic azines have a much more complex electronic absorption, and the $n \to \pi^*$ and $\pi \to \pi^*$ band overlap; being much more intense, the latter mask the former. These systems broadly resemble naphthalene in electronic absorption, as is evident in table II.

Table II   Bicyclic azines
(fine structure not given)

	λ_{max} (nm)	ϵ
quinoline	313, 270, 226	2360, 3880, 35500
quinolinium	313, 233	6350, 34700
isoquinoline	317, 266, 217	3100, 4030, 37000
isoquinolinium	331, 274, 228	4170, 1960, 37500
quinolizinium	324, 284, 225	14500, 2700, 17000
naphthalene	312, 275, 220	250, 5600, 100000

The spectra of the simple five-membered heteroaromatic systems all show just one medium-to-strong low wavelength band with no fine structure. They show no obvious similarity to the spectrum of benzene, and no detectable $n \rightarrow \pi^*$ absorption, not even in the three azoles which contain a pyridine-like azomethine nitrogen. The literature on these spectra has been confused by the reporting and discussion of spurious bands in pyrrole, thiophen, and imidazole, bands which must have been due to presence of impurities.

Table III   Five-membered heteroaromatic systems

	λ_{max} (nm)	ϵ
pyrrole	210	5100
furan	200	10000
imidazole	206	3500
oxazole	205	3900
thiophen	235	4300
thiazole	235	3000
cyclopentadiene	200, 239	10000, 3400
Derived bicyclic systems		
indole	288, 261, 219	4900, 6300, 25000
indolizine	347, 295, 238	1950, 3600, 32000
purine	263	7950

The differences between the spectra of the three related bicyclic systems, indole, indolizine, and purine, are very marked and not explicable in simple terms; the spectra of indole and indolizine, however, both bear a distant kinship with the bicyclic azines in Table II.

Nuclear Magnetic Resonance Spectra of Heteroaromatic Compounds

Although the proton chemical shifts of aromatic ring systems are not simply related to the electron density at the ring-carbons to which the hydrogens are attached, there are obvious trends in that direction: as is seen in Table IV, the lower the electron density at a particular position, the lower the τ value. Pyridine illustrates this generality quite well, with the highest τ value for the C 3 hydrogen. Another important effect is a downfield one caused by direct induction by an electronegative element, so that the C 2 hydrogen in pyridine, being much nearer to the ring nitrogen than the C 3 and C 4 hydrogens, resonates at lowest field: the same is evident in furan, and to a lesser extent in thiophen and pyrrole. Predictably, pyrylium cation with a formal positive charge shows the lowest all-round chemical shifts.

Table IV ^{1}H chemical shift values for aromatic CH

	τ_1	τ_2	τ_3	τ_4	τ_5	τ_6	τ_7	τ_8	Solvent
pyridine		1·4	3·0	2·4					CDCl$_3$
quinoline		1·1	2·5	1·7	2·1	2·4	2·3	1·9	Me$_2$CO
isoquinoline	0·7		1·5	2·3	2·1	2·3	2·3	1·9	Me$_2$CO
pyrazine		1·4							CDCl$_3$
pyrimidine		0·7		1·2	2·6				CDCl$_3$
pyridazine			0·8	2·5					CDCl$_3$
pyrylium		0·4	1·5	0·7					liq. SO$_2$
pyrrole		3·6	3·8						CDCl$_3$
thiophen		2·9	3·0						C$_6$H$_{12}$
furan		2·7	3·8						C$_6$H$_{12}$
indole		2·7	3·6	2·5	3·0	2·9	2·6		Me$_2$CO
benzofuran		2·7	3·6	2·6	2·9	2·9	2·6		Me$_2$CO
indolizine	3·7	3·4	2·9		2·2	3·7	3·5	2·8	CCl$_4$
imidazole		2·3		2·9	2·9				CDCl$_3$
thiazole		1·2		2·6	2·0				CDCl$_3$
oxazole		2·1		2·9	2·3				CCl$_4$
purine		1·5				1·3		1·7	D$_2$O
benzene	2·6								CDCl$_3$
naphthalene	2·2	2·6							CCl$_4$

The values of chemical shift are appreciably solvent-dependent, so that one must always take care to note solvents when making comparisons. The C 2 hydrogen of 4-methylpyridine may be quoted as an extreme case: this resonates at $\tau2\cdot6$ in hexane and at $\tau1\cdot4$ in DMSO.

^{13}C and ^{17}O chemical shifts of ring atoms are believed to be much more closely related to the extent to which a particular system is stabilized by aromatic resonance than are the ^{1}H chemical shifts discussed above. Such sophisticated studies are still in the exploratory phase, and we look forward with great interest to what light will be shed on the important question of aromaticity by NMR spectroscopy.

MOLECULAR ORBITAL THEORY

The semi-empirical quantum-mechanical treatment of the structure and reactivity of homo- and heteroaromatic systems in the last twenty years or so has, by making appropriate assumptions and simplifications, been giving moderate to good correlations with measured parameters such as bond-length, dipole moment, and ultra-violet absorption; more recently, *ab initio* calculations, which involve no empirical parameters, have also been giving satisfactory results.

Unfortunately the most widely quoted parameters, calculated π-electron densities, have only very recently become experimentally measurable:

values obtained by calculation have naturally varied according to the
detailed theory used and the assumptions made. Although values consistent with chemical reactivity have been obtained in most cases, this
has not been so in several. For a time, π-electron density was believed to
be linked with reactivity to electrophilic or nucleophilic reagents, but as it
became clear that this is not necessarily so, a search for a better analysis
of reactivity was made which led to calculations of polarizability, frontier
electron density, and of atom localization energy. The last parameter
gives the energy of Wheland reaction intermediates, such as 74, 75, and
76, and this gives an indication of relative positional reactivity. Indeed,

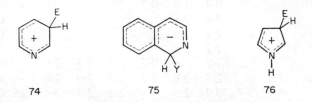

74   75   76

in the discussions of reactivity which follow in this book, qualitative
estimates of the relative stabilities of such intermediates are employed as
models for the relative stabilities of transition states leading to them.

It is probably fair to say that molecular orbital treatment of systems
with heteroatoms is still somewhat crude and beset with difficulties, and
that a discussion of the current situation is too complex to fall within
the scope of an elementary presentation of heterocyclic chemistry.

Calculated values of π-electron density and of atom-localization energy

Table V   Calculated π-electron densities

	N	2	3	4	Method	Year
Pyridine	1·14	0·98	0·98	0·95	HMO	1958
	1·20	0·92	1·00	0·95	HMO	1959
	1·10	0·95	1·01	0·98	CISCF	1957
	1·17	0·90	1·04	0·95	IDP	1963
	1·43	0·84	1·01	0·87	EHT	1965
	1·33	0·90	0·98	0·91	OHMO	1961
	1·01	1·005	1·002	0·975	*ab initio*	1967
	1·14	0·98	0·98	0·95	*ab initio*	1968
Pyrrole	0·69	1·10	1·06		HMO	1947*
	0·70	1·08	1·07		OHMO	1961
	0·66	1·075	1·095		*ab initio*	1969
	0·64	1·15	1·03		VESCF	1959
	0·77	1·06	1·05		PPP	1965
	0·71	1·08	1·07		SCMO	1967

* One of the earliest calculations.

are still subject to considerable error, so that in order not to give the student a false notion of their real value, they will not be quoted in the body of the text in this book; recent values of π-electron densities calculated for pyridine and pyrrole are given in Table V and show how dependent such values are on the type of calculation used and on the assumptions made.

2

The Synthesis of Aromatic Heterocycles

The synthesis of benzene compounds nearly always begins with an appropriately substituted and readily available benzene derivative, and only on very rare occasions has a synthesis to start with aliphatic compounds and involve the formation of the benzene ring itself. The synthesis of heterocyclic compounds presents a very different picture, for it involves ring synthesis more often than not.

When considering the synthesis of a particular heterocyclic compound, it is always important first of all to give much thought to the possibility of starting with a commercially available compound that already contains the heteroaromatic system, and then introduce and modify substituents, as in benzene chemistry; thus, the synthesis of tryptophan starts with indole (see p. 288), and the synthesis of furylpropionic acid starts with furfural (cf. p. 244). If, however, there is no obvious route from a simple and readily available heterocycle, a ring synthesis has to be designed which leads to an appropriately substituted heterocycle, one in which the substituents may be modified to give the desired product. An excellent example of this is a synthesis of vitamin B_6 (see p. 80).

This chapter discusses the principles and analyses the types of reaction frequently used in constructing an aromatic heterocycle, and also the way in which appropriate functional groups are placed in the reactants to achieve ring synthesis.

Reactions Most Frequently Used in Ring Synthesis

These involve the addition of a nucleophile to a carbonyl carbon (or the more reactive carbon of a protonated carbonyl group). When the reaction leads to C—C bond formation, then the nucleophile is the β-carbon of an enol, or of an enolate anion, or of an enamine, and the reaction is aldol in type, e.g.:

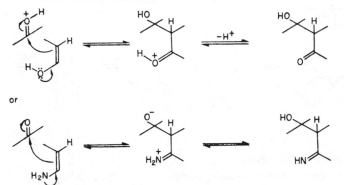

or

When the reaction leads to C-heteroatom bond formation, then the nucleophile is an appropriate heteroatom, e.g.:

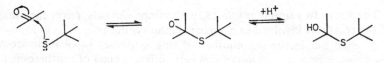

or

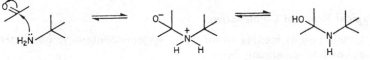

In all cases, subsequent loss of H_2O leads to the formation of double bonds, for example*

C–C bonding

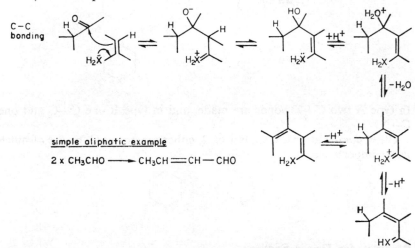

simple aliphatic example

2 x CH_3CHO ⟶ CH_3CH═CH—CHO

* In order to write generalized schemes, the abstract heteroatom X will be used, and it will be used arbitrarily in the guise of a *group 5 element*, hence as —XH_2, —X═, etc. The reader must realize, however, that the schemes also apply to *group 6 elements* Y, where —XH_2 becomes —YH and —X═ becomes —Y^+═, etc.

C—X
bonding

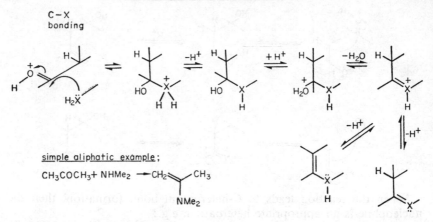

<u>simple aliphatic example</u>;

$CH_3COCH_3 + NHMe_2 \longrightarrow$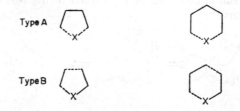

The above two simple reactions, with minor variants, cover the majority of the reactions involved in heteroaromatic synthesis.

A small but increasing number of ring syntheses involve electrocyclic reactions: a brief discussion of this very different class of syntheses is given at the end of this chapter.

Typical Reactant Combinations

Combinations leading to aromatic systems containing one heteroatom are shown schematically below:

In type A two C—X bonds are made, and in type B one C—C and one C—X.

Other combinations, such as 1 or 2, either are not used, or are of much lesser importance.

Reactants for Type A Syntheses

For five-membered rings, the use of a 1,4-dicarbonyl compound leads directly to the aromatic system:

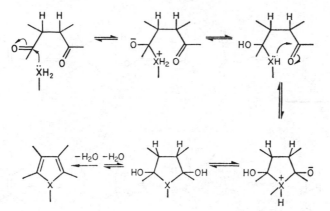

For six-membered rings, the 1,5-dicarbonyl system has to contain a
C—C double bond in order to lead directly to the aromatic system
(though synthesis can also be achieved from a saturated 1,5-dicarbonyl

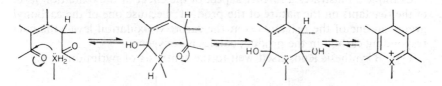

compound by way of a dihydroheterocycle, which is aromatized by
dehydrogenation, as in the Hantzsch pyridine synthesis).

Type A synthesis lends itself to the synthesis of 1,3-azoles, e.g.

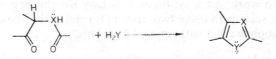

Reactants for Type B Syntheses

In this type of synthesis, since one C—C bond formation is involved,
one of the reactants must contain a nucleophilic carbon (enol, enolate
anion, or enamine): combinations of pairs of reactants such as 3 to 5
fulfil the requirements.

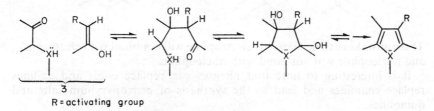

3

R = activating group

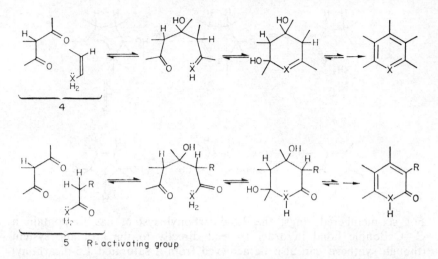

4

5   R = activating group

Example 5 illustrates a further aspect of the effect of the oxidation level of the reactants on the nature of the product: because one of the carbonyl groups in one of the reactants is at the carboxyl oxidation level, the corresponding carbon in the product carries an oxygen.

Type B synthesis lends itself well to the synthesis of pyrimidines, e.g.

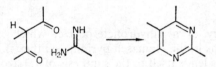

In trying to work out a synthesis it is easy for the beginner to make bad mistakes, such as to make two carbonyls react carbon to carbon, or to make a nucleophile react with an enol β-carbon.

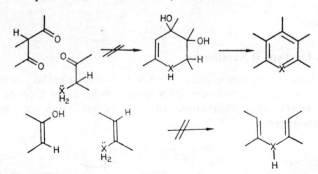

These mistakes can be avoided: electrophile will not bond with electrophile, and nucleophile will not bond with nucleophile.

It is interesting to note that phenols can replace enols and anilines replace enamines and lead to the synthesis of benzopyrylium salts and quinolines.

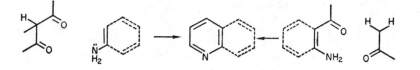

Conclusions

We can see that the chemical reactions involved in heteroaromatic synthesis are mostly simple and straightforward. The apparent 'alchemy' of a synthesis, when learned simply as reactants → products, disappears when the process is carefully analysed into a sequence of individual simple steps.

The common practice of formulating a synthesis as reactants → products is dictated in textbooks by the necessity for economy of space, and in students by the principle of speed. This is an unfortunate state of affairs which is difficult to remedy in the case of textbooks. However, in the following very close look at the synthesis of triphenylpyrylium cation, we want to show what is meant by a complete analysis of a reaction sequence. The gross reaction is:

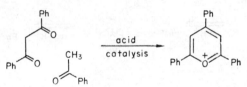

One possible sequence, 6 → 14, is as follows, and involves making the C—C bond as the first step:

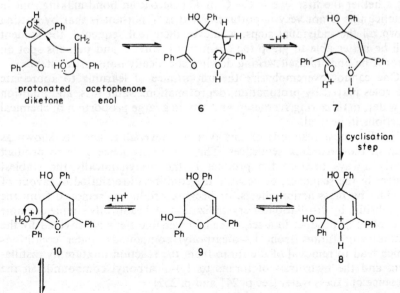

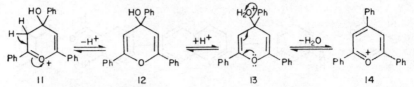

11   12   13   14

The details of this sequence can vary in the order in which the steps take place, e.g., loss of the C 6 hydroxyl in 9 could follow, instead of preceding, the loss of the C 4 hydroxyl. Another possible sequence involves making the C—O bond as the first step:

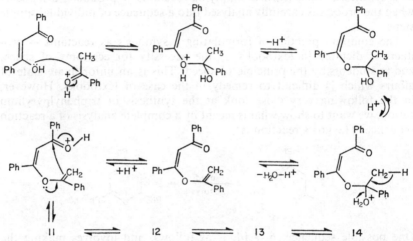

11   ⇌   12   ⇌   13   ⇌   14

In most ring syntheses of this general type there is no way of establishing whether the first step is C—C or C-heteroatom bond-making, but in practice this is not very important. What *is* important is that by putting down all the individual steps, whatever their real sequence, the student will be better able to grasp the chemistry involved, and perhaps spot an error if he finds himself writing down a chemically nonsensical step.

One cannot overemphasize the importance of learning to appreciate the roles played by protonation, deprotonation, water loss, and addition of water, in these ring syntheses, and also in a large proportion of chemical reactions in general.

Finally, most reactions of this type are reversible, and are shown as such in the preceding sequences. The overall sequence goes to product nearly always because the product is thermodynamically the stablest system in the sequence, or because the equilibria are shifted in favour of product by mass action effects, or because product is removed from the equilibrium by distillation or crystallization. A beautifully simple example of control of product in a sequence of reversible steps is provided by the synthesis of furans from 1,4-dicarbonyl compounds under conditions which lead to removal of the furan from the reaction mixture by distillation, and the hydrolysis of furans to 1,4-dicarbonyl compounds in the presence of excess water (see p. 241 and p. 252).

The Use of Electrocyclic Reactions in Ring Synthesis

The construction of a five-membered heterocyclic system by the regio-specific addition of a 1,3-dipole to a polarized multiple bond is now increasingly exploited in synthetic work. In some cases the dipolar component is an isolable species, such as a diazoalkane or an azide, but usually

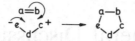

it has to be generated as a reactive intermediate in the reaction medium. 1,3-Dipolar cycloadditions have been of most value in the formation of non-aromatic systems with two or more hetero-atoms, for example in the preparation of 4,5-dihydro-isoxazoles from nitrile oxides and olefins.

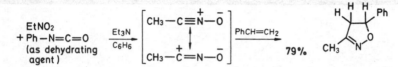

An initially formed appropriately substituted non-aromatic heterocycle may subsequently be aromatized, as the following examples show.

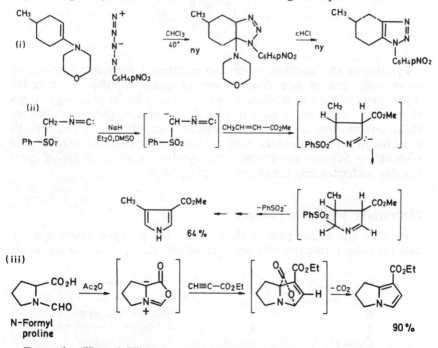

Examples (ii) and (iii) represent novel approaches to the pyrrole system: it is interesting to note that in sequence (ii) the isonitrile carbene-like carbon, although not positively charged, is electrophilic. Carbenes are *neutral* electrophiles.

3

Pyridines: General Discussion and a Comparison with Benzene

Pyridine is the simplest of the six-membered aromatic heterocyclic compounds, and as such deserves careful study. We shall see that the replacement of one CH of benzene by N leads to quite far-reaching changes in reactivity, which in the main involve decreased susceptibility to electrophilic substitution and much increased susceptibility to nucleophilic substitution when compared with benzene. Comparisons between the reactions of benzene compounds and pyridine compounds are of great value in understanding the chemistry of pyridine.

Electrophilic Addition to Nitrogen

Pyridine has a lone pair of electrons on the nitrogen which is available for bond formation with an electrophile: thus pyridine reacts easily

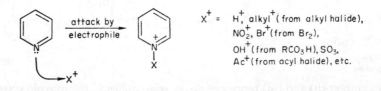

with acids to give stable protonic pyridinium salts, or with alkyl halides to give stable alkyl quaternary salts, and in such reactions behaves exactly like a tertiary aliphatic or aromatic amine. In fact, pyridine reacts with all kinds of electrophiles in this way, though only in some cases are the products stable enough to be isolated.

It is important to emphasize that these reactions occur readily only because the nitrogen lone pair is not part of the aromatic π-system (compare pyrrole, p. 9); its involvement in bonding to an electrophile does not destroy the aromaticity of the ring: pyridinium salts are still aromatic (see pp. 5 and 7).

Electrophilic Substitution at Carbon

Electrophilic substitution of pyridines at carbon is very difficult: nitration and sulphonation, for example, only take place under extreme conditions and then give only poor yields in the simple cases; Friedel-Crafts reactions fail with pyridines; less vigorous electrophilic species, which do not react with benzene anyway, have no chance of effecting pyridine C-substitution.

There are two factors responsible for this unreactivity: the first is that a pyridine ring is intrinsically less nucleophilic than a benzene ring; the second is that as soon as a pyridine compound is exposed to a medium containing electrophilic species (Hal^+, NO_2^+, or of course any acid present in the reaction medium) it forms a pyridinium salt. This increases the resistance of the ring to electrophilic attack, since now reaction necessarily leads to a high-energy doubly-charged species.

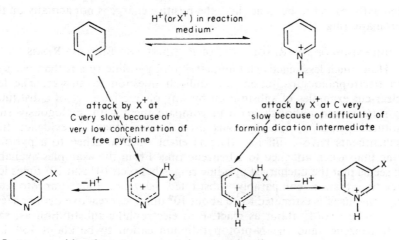

Let us consider these two aspects in more detail.

EFFECT OF ADDITION TO NITROGEN ON ELECTROPHILIC C-SUBSTITUTION

A pyridine in a reaction medium containing electrophiles is in equilibrium with pyridinium salt; the position of the equilibrium depends on the basicity of the particular pyridine and the properties and concentrations of electrophiles in the medium, but in most cases the equilibrium lies very much on the side of the salt: there is little free pyridine base present. For substitution to occur, then, there is only a choice between two unfavourable options: reaction proceeding by either free base or salt will give a slow rate of reaction, in the first case because of the very low concentration of free pyridine reactant and in the second case because of the very low reactivity of pyridinium cations.

In specific cases substitution both *via* free base or *via* conjugate acid have been observed. Thus, for example, 2,6-dichloropyridine, a relatively weak base which gives a higher than usual concentration of free base in acid solution, nitrates *via* the free base; 2,6-dimethoxypyridine, on the other hand, nitrates *via* its conjugate acid. Whether or not pyridine itself nitrates by way of its conjugate acid is not yet established.

The extent to which quaternization of a pyridine system further deactivates the nucleus towards electrophilic attack is difficult to measure, and certainly varies with the other substituents on the ring and with the particular electrophile, but the reduction in rate is of the order of 10^{13}. This can be compared with the deactivation of 10^8 for trimethylanilinium

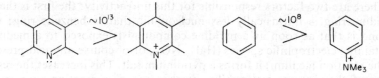

relative susceptibilities to electrophilic substitution

ion with respect to benzene; here the positive charge is not actually on the aromatic ring.

COMPARISON OF THE REACTIVITIES OF PYRIDINE AND BENZENE RINGS

How much less reactive an unquaternized pyridine ring is than benzene to electrophilic substitution is a difficult question to answer. The few clear-cut examples of substitution by way of free base are of substituted pyridines, and although one can compare these with analogously substituted benzenes, complications arise because there is evidence that substituents have a different (larger) effect when attached to a pyridine ring than when attached to a benzene ring. From the examples available it seems that the uncharged pyridine ring is between 10^3 and 10^7 times less reactive than a comparably substituted benzene: thus, unprotonated pyridine itself is estimated to be about 10^7 times less reactive than benzene —in other words about as reactive to electrophilic substitution as, say, nitrobenzene—and the N-protopyridinium cation to be about 10^{19} less reactive than benzene.

The deactivation of uncharged pyridine is ascribed to inductive and mesomeric electron withdrawal by the ring nitrogen: the mesomeric effect is expressed by the canonical structures 2 to 4. The resultant hybrid carries

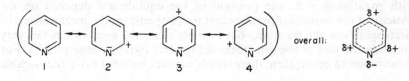

partial positive charges at the α- and γ-positions, and a partial negative charge on the nitrogen. This can be likened to the deactivation of the benzene ring in nitrobenzene, in which inductive and mesomeric electron withdrawal by the nitro group operates.

This polarization will, of course, affect both the approach of an electrophilic reagent and the energy of transition states.

POSITION OF SUBSTITUTION IN PYRIDINES AND PYRIDINIUM SALTS

Before considering pyridine specifically it is relevant to review briefly the mechanisms of electrophilic substitution. In general this can be represented as follows.

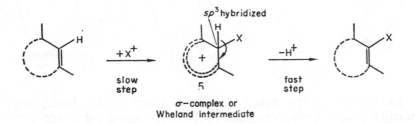

σ−complex or
Wheland intermediate

Under most circumstances such reactions are irreversible, and if isomers are formed the proportions in which they are formed reflect the different rates of reaction at the different nuclear positions.

A delocalized positively-charged addition intermediate, 5, is formed in the rate-determining step, so that relative rates of formation of these isomeric σ-complexes (Wheland intermediates) governs the observed isomer ratio. Overall substitution is completed by a rapid deprotonation of the intermediate.

Various mathematical treatments are available for rationalizing the position of attack in homo- and in hetero-aromatic systems, but these are not within the scope of this book. Instead, we shall employ throughout a valence-bond approach in which the relative stabilities of isomeric intermediates are gauged by examining the canonical forms of the structures involved. It is assumed that the relative stabilities of these intermediates, as assessed in this way, reflect the relative energies of the transition states which lead to them, and it is these energies which actually control the relative rates of reaction.

An examination of the electrophilic substitution of nitrobenzene will serve to remind us of how this approach successfully explains *meta* substitution in this case: the *ortho*, 6, and *para*, 7, intermediates are less favoured because they carry a partial positive charge on the carbon atom which is attached to the positively charged nitrogen of the nitro group; this does not occur in the *meta* intermediate, 8. Pyridines likewise undergo electrophilic substitution at a β-position, but not at α- or γ-positions, and this generalization holds for all pyridines except those which contain powerful electron-releasing substituents (see next section).

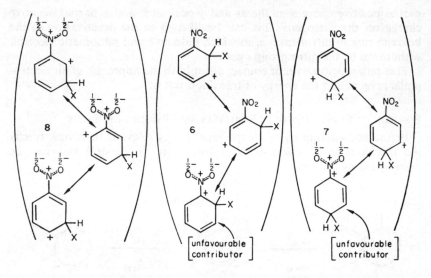

Examination of the main canonical structures of the three possible reaction intermediates shows that in the intermediates for α- and γ-substitution, 9 and 10, the delocalization of the positive charge in both cases would involve a canonical structure with a high-energy sextet positive nitrogen (of higher energy than sextet positive carbon), which can contribute but very little to stabilization: however, the intermediate of β-substitution, 11, carries the delocalized positive charge on three carbon atoms, and is thus the stablest. This intermediate is of course destabilized, in comparison with benzenonium cation, by the electron-withdrawing effect of the ring nitrogen.

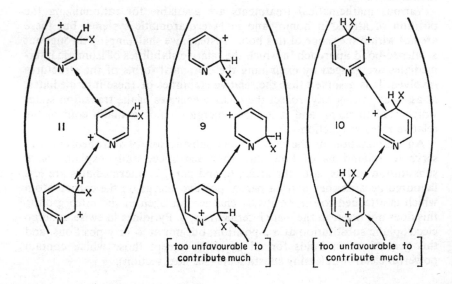

β-Substitution in a pyridinium cation finds support by a parallel argument; the very much slower rate of reaction is the consequence both of the energy required to overcome the electrostatic repulsion of two positive charges as the reactants approach each other and, of course, of the presence of two positive charges in the transition state.

EFFECT OF SUBSTITUENTS ON FURTHER SUBSTITUTION

Substituents on a pyridine ring have directing and activating/deactivating effects entirely comparable to those which they exert when attached to a benzene nucleus, though they are believed to be larger in magnitude. Qualitatively, the extent of the effect is such that one oxy- or one amino- or two alkyl substituents are sufficient to overcome the inertness of a pyridine system and allow smooth substitution, often even *via* the conjugate acid. Pyridines carrying deactivating substituents are completely inert to electrophilic substitution.

At α- or γ-position *o/p* directing groups reinforce the typical pyridine tendency for β-substitution, and thus 3 and/or 5-substitution products

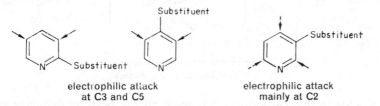

electrophilic attack
at C3 and C5

electrophilic attack
mainly at C2

result. When situated at a β-position the positive effect of the donating substituent outweighs the negative effect of the pyridine ring and substitution occurs at C 2.

The straightforward comparison with substituted benzenes falls down in some instances, for example with the 2- and 4-oxypyridines. These compounds exist (see p. 61) in the carbonyl tautomeric forms. Depending on the pH and upon the position of the oxygen, they undergo electrophilic substitution *via* O-protonated salt or *via* free pyridone, but in either

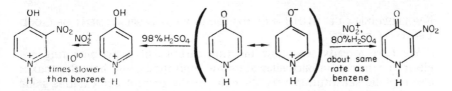

case smoothly—i.e., the pyridones are activated compared with pyridines. An analogy with the way that phenolic hydroxyl groups activate benzene substitution is clear.

Oxidation

The pyridine ring is about as resistant to oxidative breakdown as is the benzene ring. Alkaline oxidants tend to attack it more easily than neutral or acidic reagents.

Substituent alkyl groups can generally be oxidized without affecting the ring system.

Nucleophilic C-Substitution

Before dealing specifically with nucleophilic substitution in pyridines, it will be useful to recall the addition-elimination mechanism by which activated benzene compounds undergo nucleophilic substitution. For example in the ethoxydechlorination of *o*-chloronitrobenzene, the nucleo-

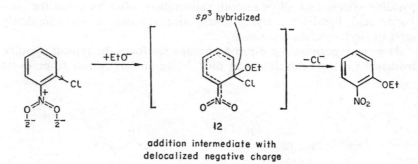

addition intermediate with
delocalized negative charge

phile adds to that carbon which is rendered δ+ both by inductive withdrawal by halogen and by inductive and mesomeric withdrawal by the nitro group, to give an anionic intermediate, 12; the overall substitution is completed by the departure of halide anion, thus regenerating a substituted aromatic system.

One can generally say, then, that aromatic nucleophilic substitution is promoted by having an electron-deficient carbon atom carrying a group capable of leaving as an anion in the second step of the process, and finally a function (nitro in the case above) capable of stabilizing negative charge in the addition intermediate.

NUCLEOPHILIC SUBSTITUTION OF PYRIDINES WITH DISPLACEMENT OF GOOD LEAVING GROUPS

We have already seen how the ring carbons of a pyridine ring are electron deficient, in particular at α- and γ-positions, and how this makes electrophilic substitution very difficult. In the present context this same polarization, in combination with a suitable leaving group (say halogen or nitro) makes for easy substitution by nucleophiles (such as amines, alkoxides, and water).

POSITION OF NUCLEOPHILIC SUBSTITUTION IN PYRIDINES

An examination of the canonical forms of the intermediates for α- and γ- versus β- substitution makes it clear that only in the two former cases can the

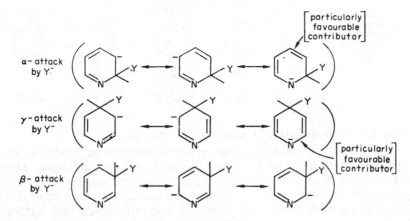

ring nitrogen play a mesomeric role in stabilizing negative charge in the

intermediate $\left[\ \text{>N:}^-\ \text{is much stabler than}\ \text{>C:}^-\right]$

Thus, although substitution of β-chloropyridine is faster than that of chlorobenzene, it is much slower than substitution at α- or γ-positions, as it is activated only by induction.

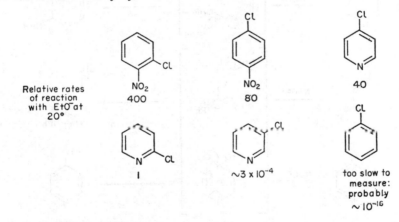

The figures above show that for attack by a negatively charged nucleophile, a γ-position is slightly more reactive than an α-position. The difference is small however, and other factors, for example the quaternization of nitrogen, can cause a reversal in the relative order.

EFFECT OF QUATERNIZATION ON NUCLEOPHILIC SUBSTITUTION

Quaternization greatly enhances the overall rate of nucleophilic displacement, essentially by increasing the rate of the nucleophilic addition step. The effect is electrostatic in nature and is felt more at an α-position (10^{13} enhancement) than at a γ-position (10^{8} enhancement).

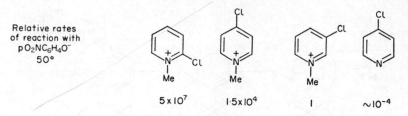

Relative rates
of reaction with
$pO_2NC_6H_4O^-$
50°

5×10^7 $1 \cdot 5 \times 10^4$ 1 $\sim 10^{-4}$

ANALOGY WITH ALIPHATIC CARBONYL COMPOUNDS

It is instructive to compare nucleophilic substitution at pyridine α- and γ-positions with nucleophilic substitutions of aliphatic analogues. Thus, at each stage of reaction, the displacement of halide from pyridine α- and γ-positions is mechanistically closely similar to the displacement of halide respectively from acid chlorides and α,β-unsaturated β-chloro ketones.

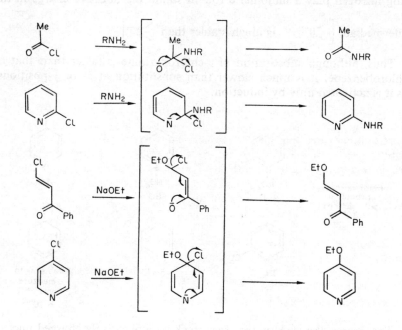

NUCLEOPHILIC SUBSTITUTION WITH DISPLACEMENT OF HYDROGEN

Substitution of pyridines with displacement of hydrogen, formally as hydride anion, is a process with very few analogies in benzene chemistry. It is observed only in pyridines at α- or γ-positions, usually at the former. It is considerably more difficult than displacement of halogen, or similar group, both because the δ+ charge on the carbon attacked is not as high and because hydride as such is a very poor leaving group. Thus, more reactive nucleophiles such as amide or alkyl or aryl anions (as in organolithium reagents) are required for the addition step and, what is more, a molecule or ion to which hydride can be transferred has to be present. This can be provided in the form of an oxidant, air, for example, or nitrobenzene in aminations; heating also causes aromatization of the intermediates. Quaternization of the pyridine ring facilitates nucleophilic addition, but an oxidant must be provided to regenerate the quaternary pyridinium aromatic system.

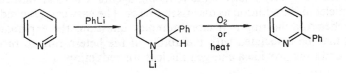

Again, a comparison with aliphatic carbonyl compounds is profitable; the addition of organolithium reagents to α, β-unsaturated ketones leads to predominant attack at the carbonyl carbon and not at the conjugated β-carbon: initial complexing at oxygen is invoked to explain this phenomenon, and a similar explanation probably applies to the preferred α attack in the pyridine cases.

Free Radical Substitution

Pyridines, pyridinium ions, benzene and nitrobenzene all react with free radicals at about the same rate: the vast rate differences observed in electrophilic substitution, and to a lesser extent in nucleophilic substitution, are not in evidence, and this makes sense, for the free radical reagents are of course uncharged. Very reactive radicals, such as Ph·, differentiate poorly between the three positions: less reactive radicals in general lead to 2- and 4-substitution products in preference to 3-substitution products, and this differentiation increases, as does rate of reaction, in acidic solution.

Homolytic substitution of aromatic compounds proceeds by radical addition followed by hydrogen atom abstraction. The positional selectivity then is seen as reflecting a greater stabilization of radicals formed by attack at α or γ-positions: once again, involvement of the nitrogen in the resonance is only possible for α and γ attack.

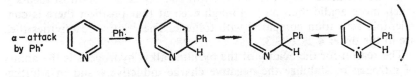

etc. for β and γ attack

In the radical phenylation of nitrobenzene, the observed o/p substitution reflects the potential for delocalization involving the nitro group in addition intermediates.

Reduction

Generally speaking, pyridine rings are more easily reduced than benzene rings. Pyridines are easily reduced catalytically to piperidines; partial reduction to tetrahydro or dihydro derivatives results from sodium-ethanol, lithium aluminium hydride or sodium-ammonia treatments; metal-acid combinations are without effect on pyridine.

For the obviously nucleophilic reducing agents it is clear that reaction is by nucleophilic addition as discussed earlier. The easy catalytic reduction of the heterocyclic ring is a more complex question, the explanation of which may be associated with the ability of the hetero-ring to protonate easily and thus provide a charged nucleus for reduction.

Properties of Alkyl Groups on Pyridine Rings

Alkyl groups attached to pyridine rings are much more reactive than when attached to benzene rings, in the sense that the alkyl hydrogen atoms are acidic. Condensation reactions of the anions thus derived are of great value and have been much used in pyridine chemistry.

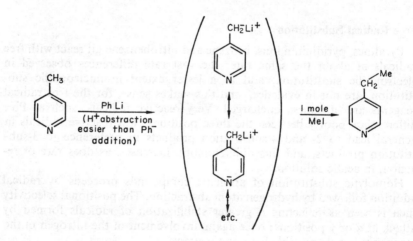

In the isomeric alkyl pyridines, hydrogen on α- and γ-substituents is much more acidic than on β, though even at that position there is considerable activation compared with toluene. The relative reactivities then are in the order $4 > 2 \gg 3$.

The reason for the acidity of the pyridine alkyl hydrogens is the ability of nitrogen to stabilize the negative charge inductively and in addition mesomerically at the α and γ positions. Once again it is fruitful to draw an analogy with aliphatic compounds, for which these processes parallel enolate formation in ketones and conjugated ketones.

Quaternization of nitrogen renders the alkyl hydrogen atoms even more acidic, for example, at room temperature for the methosalts of the

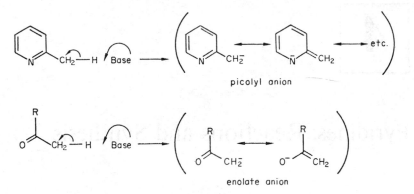

picolyl anion

enolate anion

methyl pyridines the α-isomer exchanges in water-potassium carbonate the γ-isomer in water-sodium hydroxide, conditions which do not affect the β-isomer. N-Quaternization once again has a greater effect at the α-position, as we saw in nucleophilic substitution. Deprotonation of the salts gives reactive enamines, and is the counterpart of the formation of enols in aliphatic chemistry.

4

Pyridines: Reactions and Synthesis

The simple pyridines are stable and relatively unreactive liquids, with strong penetrating odours that are very unpleasant to some people and quite acceptable to others. They are much used as solvents, especially pyridine itself, and also as bases in a variety of reactions, the most familiar of which is tosylation. Pyridine and the monomethyl pyridines (the picolines) are completely miscible with water.

Pyridine was first isolated, like pyrrole, from bone pyrolysates: the word is constructed from the Greek for fire, which is *pyr*, and the suffix *idine* which was at the time being generally applied to aromatic bases, e.g., toluidine, phenetidine, etc.

Pyridine and the simple alkyl derivatives were for a long time produced from coal tar, in which they occur in quantity. This source is now being displaced by synthetic processes: pyridine itself, for example, can be produced commercially in 60–70 per cent yields by the gas phase interaction of crotonaldehyde, formaldehyde, steam, air and ammonia on silica-alumina catalysts at 400°C; other commercial synthetic processes lead to mixtures of pyridine and 3-methylpyridine.

The aromatic pyridine ring plays a very important role in fundamental metabolism in two ways: as an oxidizing system by effective hydride abstraction in nicotinamide adenine dinucleotide (NAD^+) in dehydrogenase enzymes (see p. 69), and in transamination reactions, an important aspect of amino acid metabolism, as vitamin B_6 (see p. 81).

Desmosine is an amino acid characteristic of the protein elastin, which possesses unique elasticity and tensile strength.

Nicotine, a pyridine alkaloid, is a secondary metabolite of the familiar tobacco plant. In synthetic chemotherapeutics the pyridine ring turns up in isoniazide, which is used in the treatment of tuberculosis, and in sulphapyridine, one of the famous sulphonamide antibacterials. More recently, paraquat has become familiar in gardening as a weed-killer.

Isoniazide

Sulphapyridine

Nicotine

Paraquat

Nicotinamide adenine
dinucleotide (NAD⁺)

Desmosine

REACTIONS AND SYNTHESIS OF PYRIDINES

A   Reactions with Electrophilic Reagents

1   ADDITION TO NITROGEN

In reactions which involve bond formation with the lone pair of electrons on the ring nitrogen, such as protonation and quaternization, pyridines behave just like tertiary aliphatic or aromatic amines.

When a pyridine reacts as a nucleophile in this way it gives a pyridinium cation in which the aromatic sextet is retained, and the nitrogen acquires a formal positive charge.

(a) Protonation of nitrogen.   Pyridines form crystalline salts with most protic acids. Pyridine itself, with pK_a 5·2 in water, is a much weaker base than saturated aliphatic amines which have pK_a values mostly between 9 and 11. Since the *gas phase* basicity of pyridine has been found to be actually of the same order of magnitude as that of the aliphatic amines, the pK_a values observed in aqueous solution would seem to reflect very strong solvation effects which, for reasons which are not yet clear, stabilize aliphatic ammonium cations very much more than pyridinium cations. The argument hitherto used to rationalize the low pK_a of pyridine in water was based on the stabilization of the electron lone pair on nitrogen

by the greater contribution of the *s* component in an *sp*² hybridized orbital when compared with *sp*³, and found support in the effectively non-basic character of the *sp* lone pair in nitriles: the gas phase measurements seem to cast doubt on the validity of such arguments.

Electron-releasing substituents such as methyl generally increase the basic strength of the pyridine to which they are attached, but with other classes of substituent the situation becomes too complicated to be discussed here.

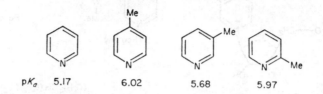

pK_a 5.17   6.02   5.68   5.97

(*b*) *Nitration of nitrogen.* This occurs very readily by interaction of pyridines with nitronium salts such as nitronium borofluoride. Protic nitrating agents such as nitric acid of course lead exclusively to N-protonation.

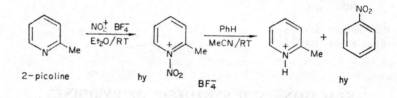

2-picoline   hy   hy

1-Nitro-2-methylpyridinium borofluoride has been used as a non-acidic nitrating agent: the 2-methyl compound is used because in it the 2-methyl group sterically inhibits stabilizing resonance overlap between the nitro group and ring, and is in consequence a more reactive nitronium ion donor.

(*c*) *Sulphonation of nitrogen.* Pyridine reacts with SO_3 to give the crystalline zwitterionic pyridinium-1-sulphonate, usually referred to as the pyridine-sulphur trioxide compound.

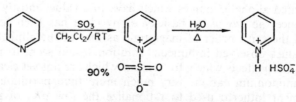

                           90%

pyridinium-1-sulphonate

This compound is quite reactive: it is hydrolysed in hot water to sulphuric acid and pyridine, and it has found much use as a mild sulphonating agent (see p. 196 and p. 260; for reaction with OH⁻, see p. 71).

(*d*) *Halogenation of nitrogen.*   Pyridines react easily with halogens and interhalogens to give crystalline compounds best formulated as resonance hybrids formally related to the tribromide and tri-iodide anions. They are largely undissociated in non-polar solvents such as CCl_4.

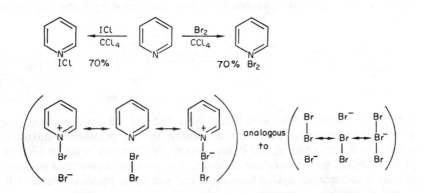

It is important to distinguish these compounds from the crystalline 'pyridine hydrobromide perbromide', which is obtained by treating pyridine hydrobromide with bromine, and which does not contain an N-

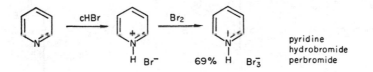

halogen bond: this salt can act as a source of molecular bromine and is useful when small and accurately-known weights of bromine are required.

(*e*) *Acylation of nitrogen.*   Carbonyl halides react very rapidly to give 1-acylpyridinium salts, which are themselves very reactive and widely used

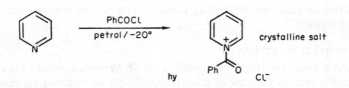

acylating agents. The usual acylation procedure is to add the alcohol or amine to a solution prepared by the addition of the acyl halide to an excess of pyridine. This is useful in the acylation of acid-labile molecules.

(*f*) *Alkylation of nitrogen.* Alkyl halides and sulphates react readily with pyridines to give quaternary pyridinium salts. As with aliphatic tertiary amines, increasing substitution around the nitrogen or around the halogen- or sulphate-bearing carbon causes an increase in the alternative, competing, elimination reaction which gives olefin and tertiary salt: thus collidine (2,4,6-trimethylpyridine) is often used in dehydrohalogenation reactions.

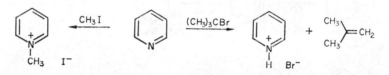

Pyridines react with acetylenes activated by electron-withdrawing groups. The products can vary according to the reaction conditions, solvent, and temperature, but in all cases these reactions begin by electrophilic addition to the pyridine nitrogen. The process is rendered irreversible by a subsequent cyclization step. For example, pyridine reacts with dimethyl acetylene dicarboxylate by the path indicated here: the cyclization

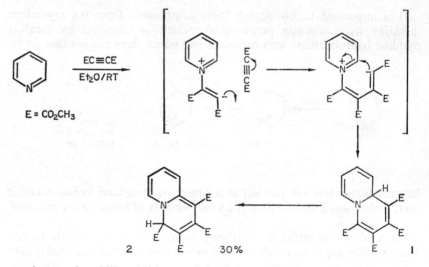

step is a nucleophilic addition to C 2 of the pyridinium cation (see p. 68); the initial bicyclic compound, 1, tautomerizes to the stabler, more conjugated isomer, 2.

2   SUBSTITUTION AT CARBON

In most cases electrophilic substitution of pyridines occurs very much less readily than in the case of a correspondingly substituted benzene. The main reason for this big difference is that the electrophilic reagent reacts preferentially with the pyridine nitrogen to generate a pyridinium cation (not necessarily only 1-protopyridinium, as is seen in the previous section): this cation will naturally be very resistant to further attack by electrophilic reagent.

When it occurs, then, electrophilic substitution at carbon may represent either highly-unfavoured attack of a pyridinium cation or much easier attack of a very low equilibrium concentration of uncharged free pyridine base. This type of situation is typical of the reactions of electrophilic reagents with bases in general, a well-known instance being the diazotization of aniline.

Some of the typical electrophilic substitution reactions do not occur at all: Friedel-Crafts acylation and alkylation, and the Hoesch reaction fall into this category, but then these also fail with nitrobenzene and with acetophenone.

Milder electrophilic reagents such as the Mannich reagent, diazonium salts, nitrous acid, etc., which do not react with benzene, likewise fail to effect C-substitution in pyridine.

One ingenious experiment, the reaction of SO_3 with 2,6-di*tert*butyl-pyridine (see p. 50), gives a clue to the reactivity of the neutral pyridine ring: it shows it to be indeed less reactive than the benzene ring, but *very much* more reactive than a pyridinium cation.

(*a*) *Protonation.* H-D exchange by electrophilic addition to carbon in neutral pyridine or pyridinium cation (by the type of mechanism which operates in the acid-catalysed H-D exchange of benzene) has not been observed. H-D exchange at C 2 and C 6, however, does occur in DCl-D_2O, but it occurs by C 2 (or C 6) deprotonation of pyridinium cation (see p. 57).

(*b*) *Nitration.* Pyridine is very difficult to nitrate on carbon: even under extreme conditions only a small yield of 3-nitropyridine is obtained.

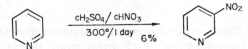

Alkyl substitution facilitates nitration, but two methyl groups are required to provide sufficient activation for substitution to compete effectively with oxidation of the side-chain. That 3-nitro-2,4,6-trimethylpyridine is formed by nitration of the 1-proto-2,4,6-trimethylpyridinium cation and

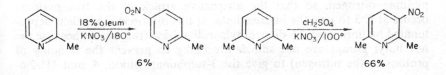

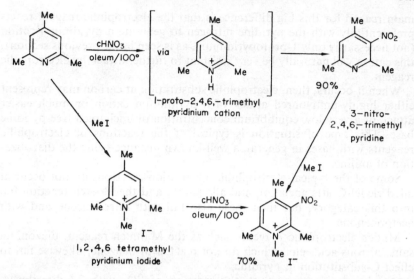

not the conjugate base is shown by the nitration of 1,2,4,6-tetramethyl-pyridinium cation under the same reaction conditions.

(c) *Sulphonation.* Pyridine is very resistant to sulphonation by conc. H_2SO_4 or oleum, only very low yields of pyridine-3-sulphonic acid being produced after prolonged reaction periods at 320°. A catalytic quantity of $HgSO_4$, however, results in smooth sulphonation either in conc. H_2SO_4 or in oleum. The role of the catalyst is not clearly understood, but it

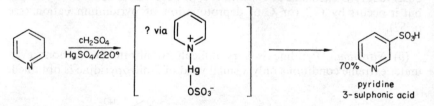

seems probable that a pyridine-mercuric sulphate compound is involved, which allows back-donation of electrons from the Hg—N bond to stabilize the transition state for the substitution.

One of the very few instances in which electrophilic substitution has been achieved on a simple uncharged pyridine ring, that is not as in the great majority of cases on a pyridinium cation, is the sulphonation of 2,6-di*tert*butylpyridine by SO_3. In this pyridine base, the bulky *tert* butyl groups sterically prevent the large SO_3 molecule from bonding with the pyridine nitrogen, so that the alternative attack of the free pyridine occurs at C 3 to give the intermediate, 3; this intermediate is then depro-tonated by unreacted 2,6-di*tert*butylpyridine base (this occurs because the *tert* butyl groups are not sufficiently bulky to prevent the bonding of protons to the nitrogen) to give the 3-sulphonate anion, 4, and 1H-2,6-

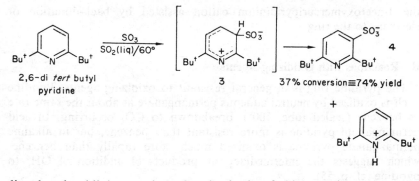

2,6-di *tert* butyl
pyridine

3

37% conversion≡74% yield

di*tert*butylpyridinium cation. On the basis of this experiment one can deduce (by a series of comparisons which we need not detail here) that pyridine base, as distinct from pyridinium cation, is approximately as susceptible to electrophilic substitution as nitrobenzene.

(*d*) *Halogenation.* Pyridine reacts with chlorine and bromine at C 3 and C 5. Catalysts such as pumice aid the reaction. 3-Bromopyridine is

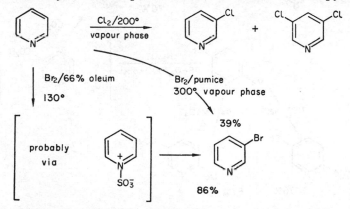

produced in good yield by the action of bromine in oleum: this substitution is thought to proceed *via* pyridinium-1-sulphonate, since no bromination occurs in 95 per cent sulphuric acid under comparable conditions. 3-Chloro pyridine is produced by chlorination at 100° in the presence of aluminium chloride.

(*e*) *Acetoxymercuration.* The salt formed by the interaction of pyridine and mercuric acetate at RT can be rearranged to 3-acetoxymercuripyridine by heating to a relatively moderate temperature. This process, like that of the $HgSO_4$-catalysed sulphonation, may involve electrophilic attack on

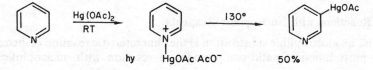

the 1-acetoxymercuripyridinium cation assisted by back-donation of electrons to the ring.

B   Reactions with Oxidizing Agents

The pyridine ring is in general resistant to oxidizing agents. Pyridine itself is oxidized by neutral aqueous permanganate at about the same rate as benzene (sealed tube, 100°), breakdown to CO_2 occurring. In acid permanganate pyridine is more resistant than benzene, but in alkaline permanganate pyridine is oxidized much more rapidly than benzene, which suggests the intermediacy of products of addition of OH^- to pyridine (cf. p. 55).

In most instances substituent groups may be oxidized with survival of the pyridine ring, thus alkyl pyridines are oxidized to pyridinecarboxylic acids.

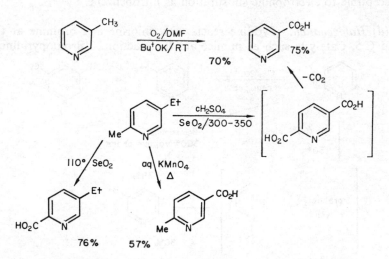

In common with other tertiary amines, pyridines react smoothly with percarboxylic acids to give N-oxides. Pyridine N-oxides provide much interesting chemistry and are very useful synthetic intermediates (see p. 70).

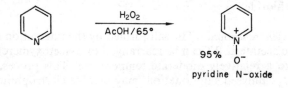

pyridine N-oxide

C   Reactions with Nucleophilic Reagents

Just as electrophilic substitution is the characteristic reaction of benzene and most homoaromatic compounds, so reaction with nucleophilic reagents is characteristic of pyridines. In this they resemble carbonyl compounds (see p. 40).

It is very necessary to realize that nucleophilic substitution of hydrogen differs in an important way from electrophilic substitution: whereas the last step in electrophilic substitution is loss of proton, an easy process, the last step in nucleophilic substitution of hydrogen has to be a hydride

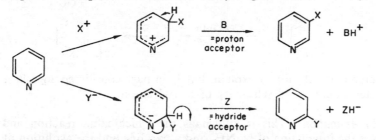

transfer, which is less straightforward and generally needs the presence of an oxidizing agent as hydride acceptor. Nucleophilic substitution of an atom or group which is a good anionic leaving group, however, is an easy and straightforward process.

1   SUBSTITUTION WITH HYDRIDE TRANSFER

(a) *Alkylation and arylation.* Reaction with alkyl or aryllithium compounds proceeds in two discrete steps: addition of the organolithium compound occurs very readily to give a dihydropyridine lithium salt which can in some cases be isolated: this may subsequently be oxidized to

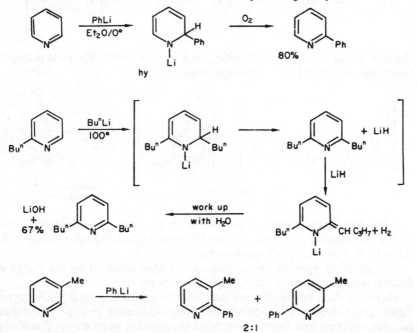

the aromatic pyridine, either with an added oxidizing agent, or by disproportionation, or thermally by loss of lithium hydride (which subsequently deprotonates an α– or γ-alkyl group if present, as shown above).

Almost always, reactions of this type have been observed to occur at an
α-position: however, benzyllithium reacts with pyridine at C 4. The

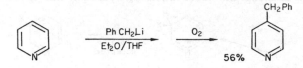

difference is difficult to explain, but is in part considered to reflect the
greater stability and selectivity of benzyl anion.

(b) *Amination.* This is known as the Chichibabin reaction and involves the interaction of $NaNH_2$ and a pyridine and the evolution of H_2.
In simple cases α-substitution occurs, but where both α-positions are
blocked it is possible to obtain a γ-aminopyridine, though in poorer yield.
The exact nature of the mechanism is not known, but it must involve
hydride transfer to a proton donor to generate the molecular hydrogen.
Once a little 2-aminopyridine has been formed, the reaction could be a

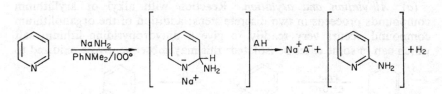

chain in which the 2-aminopyridine is the proton donor, the final product
being the sodium salt of 2-aminopyridine.

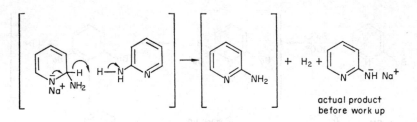

It has been argued that the dominance of α- against γ-amination is due
to hydride transfer occurring in a cyclic system involving association with
the ring-N, but this is still quite speculative.

Amination of pyridines with sodamide is also achieved by the use of a
hydride acceptor such as NO_3^- in molar proportions.

More vigorous conditions are necessary to aminate α- and γ-alkylpyri-
dines, since proton abstraction from the side-chain (see p. 66) is the
favoured process and substitution must presumably proceed via dianionic

species. 2-Aminopyridine is converted into 2,6-diaminopyridine by $NaNH_2$ in $PhNMe_2$ at 170°, the intermediate here may be 5.

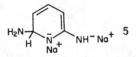

(c) *Hydroxylation.* Hydroxide ion being a much weaker nucleophile, attacks pyridine only at very high temperatures, to give a poor yield of the sodium salt of α-pyridone. The corresponding reaction with isoquinoline proceeds much more smoothly (see p. 111).

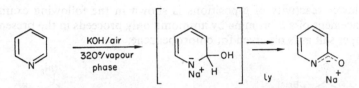

(d) *Addition of bisulphite.* Pyridine reacts with sodium bisulphite by *addition* to give piperidine 2,4,6-trisulphonate. The first step here is very

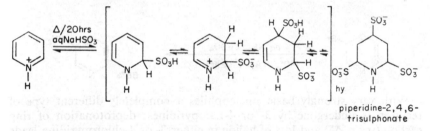

likely to be nucleophilic addition to pyridinium cation (cf. pyrrole p. 195).

2   SUBSTITUTION WITH DISPLACEMENT OF HALIDE OR NITRITE

Halogen and nitro substituents at α or γ-positions, but not at β-positions, are easily displaced by a wide range of nucleophiles: the γ-halopyridines are more reactive than the α-isomers. In their relative resistance to nucleophilic displacement, β-halopyridines resemble halobenzenes.

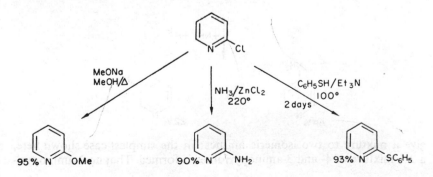

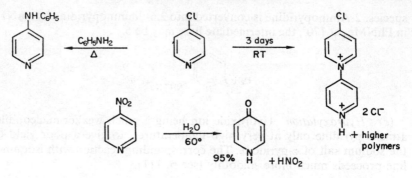

The lower reactivity of β-positions is shown in the following examples: displacement of a β-bromine by ammonia only proceeds in the presence of a copper salt as is observed for chlorobenzene.

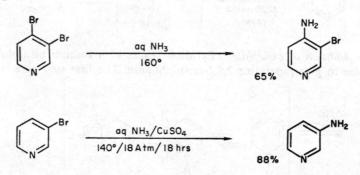

With very strongly basic nucleophiles a completely different type of reaction is undergone by 3- or 4-halopyridines: deprotonation of ring carbon (see p. 57) and loss of halide in either 3- or 4-chloropyridine leads to the same transient 3,4-pyridyne which then adds NH_2^- and proton to

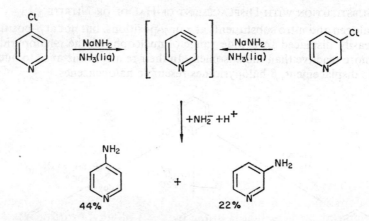

give a mixture to two isomeric amines: in the simplest case shown here, a 2:1 mixture of 4- and 3-aminopyridines is formed. That no 2-aminopyridine is produced from 3-chloropyridine and that the *same* ratio of 3- to 4-aminopyridine is produced from both 3- and 4-chloropyridine suggests that under these conditions the isomeric 2,3-pyridyne is not formed: 2,3-pyridyne can however be generated from 3-bromo-2-chloropyridine.

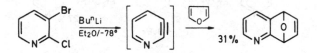

3   DEPROTONATION OF RING CARBON

When pyridine is heated to about 200° in D_2O-NaOD, complete H—D exchange occurs: this is believed to occur by way of deprotonation to yield very short-lived anions which then pick up H^+ or D^+ from the solvent.

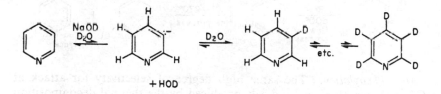

Deuteration at only α-carbon atoms is observed in neutral or acidic media. Exchange in these cases involves deprotonation of the 1-proto-pyridinium cation to give specifically a 1,2-ylid; 1,3- and 1,4-ylids evidently are too unstable to be formed.

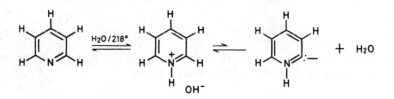

4   REACTION OF HALOPYRIDINES WITH LITHIUM, MAGNESIUM, AND LITHIUM ALKYLS. METALLOPYRIDINES

Halopyridines can only be induced to react with magnesium by the entrainment method, that is in the presence of ethyl bromide. Halopyridines cannot be made to react with lithium metal to give lithium pyridyls: these, however, are conveniently prepared in good yields by halogen-metal

interconversion with butyllithium, and are useful synthetic intermediates which have largely displaced pyridyl Grignards.

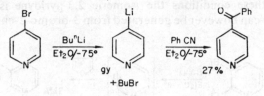

D   Free Radical Reactions

(a) *Halogenation.* At temperatures where Br_2 and Cl_2 molecules are
appreciably dissociated into atoms, the products of reaction with pyridine
are different from those obtained by electrophilic substitution at lower
temperatures. These 2- and 2,6-disubstituted products are thought to
arise by attack of halogen atoms on the ring.

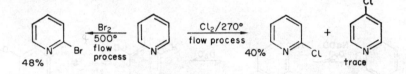

(b) *Phenylation.* The same high degree of selectivity for attack at
C 2 is shown by phenyl radicals produced by the thermal decomposition
of benzoyl peroxide. Phenyl radicals produced by other routes are much

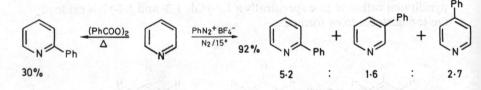

less selective and lead to all three phenylpyridines, as for example with
the decomposition of $PhN_2{}^+BF_4{}^-$ shown above.

(c) *Alkylation.* Tertiary alkyl radicals generated in the presence of
1-protopyridinium cations display their nucleophilic character in high
regioselectivity, alkylating at C 2; secondary and primary radicals are
however less selective and give 2- and 4-monosubstitution and 2,4-disubsti-
tution.

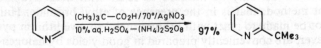

(*d*) *Dimerization*.   Raney nickel catalyses the production of 2,2'-dipyridyl from pyridine. The reaction is considered to proceed by bonding of pyridine to a nickel atom to form a mesomeric radical which then dimerizes to an intermediate which breaks down to H_2, nickel, and 2,2'-dipyridyl.

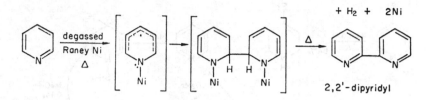

2,2'-dipyridyl

In an analogous reaction, sodium in THF or HMPA surprisingly effects dehydrogenation to 4,4'-dipyridyl and sodium hydride.

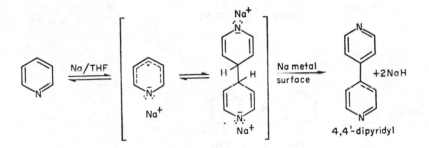

4,4'-dipyridyl

In recent years these dipyridyls have gained importance as intermediates in the production of weed-killers such as paraquat, which is 1,1'-dimethyl-4,4'-dipyridinium chloride.

E   Reactions with Reducing Agents

Pyridines are much more readily reduced than benzenes. Catalytic reduction usually proceeds to completion at atmospheric pressure and ambient temperatures.

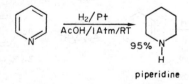

piperidine

Sodium-ethanol reductions usually give mixtures in which the 1,2,5,6-tetrahydroisomer is the main product. Sodium in liquid ammonia-ethanol, on the other hand, reduces to give highly reactive 1,4-dihydropyridines. 1,4-Dihydropyridine itself can be produced indirectly in two ways, either by trimethylsilane reduction and then removal of the N-trimethylsilyl group or by sodium borohydride reduction of 1-methoxycarbonyl-pyridinium chloride produced *in situ*, again with subsequent removal of the N-protecting group.

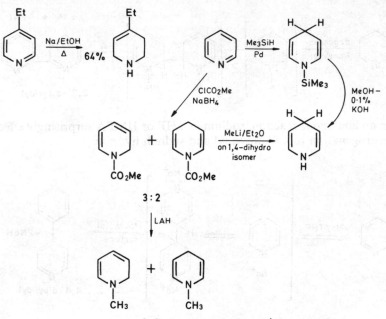

3 : 2

3 : 2 changed to 8 : 92 by Bu^{t}OK / DMSO / 92°

N-Methyl-1,4-dihydropyridine is the dominant tautomer when an equilibrium is set up between this and its isomer.

The action of acetic anhydride-zinc on pyridine itself leads to a dimeric product (6); 4-alkylpyridines however give monomeric products, with a 1,4-diacetyl-1,4-dihydro-4-alkylpyridine structure.

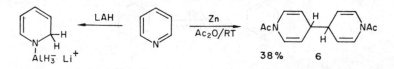

38%   6

Sodium borohydride does not reduce pyridines, though it reduces pyridinium salts (see p. 68); the more powerful LAH adds to give a 1,2-adduct which is too reactive to be isolated.

Rather surprisingly, metal-acid combinations are without effect on pyridines: surprisingly, because the C=N$^+$ function is usually readily reduced (see p. 205).

F Oxy- and Aminopyridines

1 STRUCTURE

The three oxypyridines are subject to tautomerism involving H interchange between O and N, and here again we see a big difference between the β-isomer on the one hand and the α- and γ-isomers on the other.

β-Hydroxypyridine exists in equilibrium with a corresponding zwitterionic tautomer, the exact ratio varying with solvent. The zwitterionic species is best viewed as a resonance hybrid to which 7 makes an important contribution (see also p. 64).

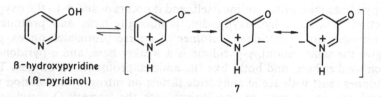

β-hydroxypyridine
(β-pyridinol)

7

In all solvents, except petrol at high dilution, the 2- and 4-hydroxypyridines are extremely unfavoured in the equilibrium: polar solvation effects strongly favour the amide-like structures of the α- and γ-pyridone

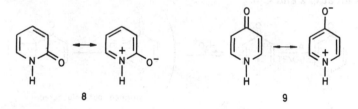

8   9

tautomers, 8 and 9. However, in petrol solution in the absence of effective solvation and at concentrations of less than 10^{-5} molar to minimize dimeric hydrogen-bonded association, the 2-hydroxypyridine:α-pyridone ratio is around 2:3, changing even further to 2:1 in the gas phase: this shows the unsolvated tautomers to be of approximately equal thermodynamic stability.

To sum up then, β-hydroxypyridine exists as such in equilibrium with the zwitterionic tautomer, and the other two isomers exist in solution as α- and γ-pyridones, with highly polarized amide-type structures.

The situation with the aminopyridines is that all three exist as amino tautomers: α- and γ-aminopyridines, 10 and 11, are polarized in a sense opposite to that in the pyridones.

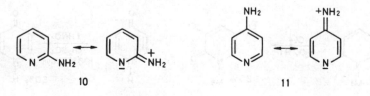

10   11

2   REACTIONS OF PYRIDONES

(a) *Electrophilic reagents.*   As might be expected, electrophilic substitution at carbon occurs very much more readily with the three

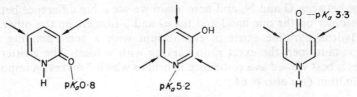

oxypyridines than with pyridine itself, and it occurs *o-* and *p-* to the oxygen function. Whereas β-hydroxypyridine is normally basic and protonates on nitrogen (its zwitterionic tautomer of course protonates on oxygen to give the same cation), γ-pyridone is a weaker base, and α-pyridone a much weaker base, and both, like the amides, protonate on oxygen. The pyridones react with acetic anhydride fastest on nitrogen; in solution the N-acyl derivatives come to equilibrium with the isomeric O-acetyl compounds. The 2-acetoxy isomer predominates by 9:1, but 1-acetyl-4-pyridone is as stable as 4-acetoxypyridine.

Acid-catalysed H—D exchange in γ-pyridone occurs at C 3 and C 5 and not at all at C 2 and C 6.

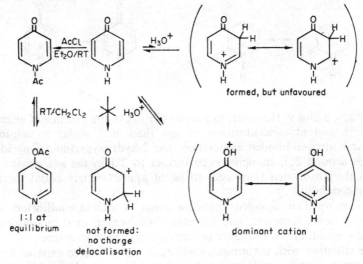

α- And γ-pyridones are readily halogenated, nitrated, and sulphonated. γ-Pyridone nitrates at about the same rate as 4-methoxypyridine, indi-

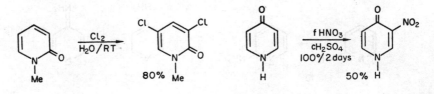

cating that electrophilic attack is occurring on the O-protonated cation, however α-pyridone, which can be nitrated at either C 3 or C 5, reacts in both cases *via* the free base. Even so, selectivity relies on a choice of more (for C 5) or less (for C 3) acidic conditions and must thus depend on selective solvation effects.

3-Hydroxypyridine and the 2-, 3-, and 4-alkoxy pyridines likewise undergo electrophilic substitution easily and at positions governed by the directing effect of the oxygen substitutent.

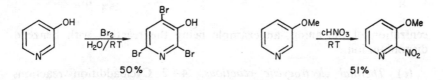

In α- and γ-pyridones, electrophilic addition to oxygen can trigger off an easy nucleophilic attack of the resulting cation. This is seen in the

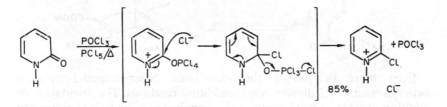

synthetically important conversion of pyridones into the corresponding chloropyridines.

(*b*) *Deprotonation.* Pyridones are appreciably acidic and are readily deprotonated to give mesomeric anions. The canonical forms 12 and 13, in

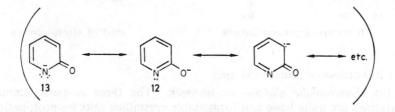

which the negative charge is on O and N, are the main contributors to the mesomeric anion. These anions are good nucleophiles and easily alkylated

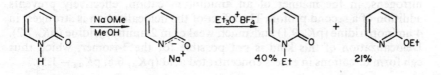

and acylated at both oxygen and nitrogen, the ratio of products depending on the solvent, counter-ion, and reagent.

Some pyridone reactions proceed by way of a low equilibrium con-

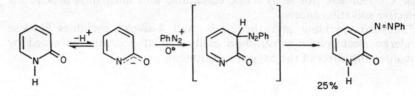

centration of the anion, an example being the reaction with benzene diazonium ion.

(c) *Thermal electrocyclic reactions.* 4+2 Cycloaddition reactions have been demonstrated for α-pyridones.

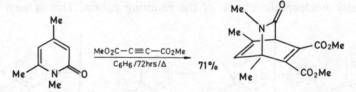

Even more intriguingly the meso-ionic N-substituted-3-pyridinol betaines react as 1,3-dipoles in cycloaddition reactions. The orientation of addition provides a good guide to the significance of polarized resonance contributors such as 14.

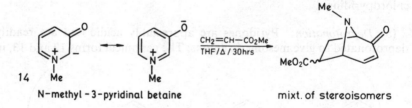

14　Me　　　　　　　　Me

N-methyl-3-pyridinal betaine　　　　　　　mixt. of stereoisomers

3　REACTIONS OF AMINOPYRIDINES

(a) *Electrophilic addition to nitrogen.* The three isomeric amino-pyridines are quite basic and form stable crystalline salts by protonation of the ring nitrogen. Here again there is a sharp difference between the α- and γ-isomers on the one hand and the β-isomer on the other: the α- and γ-amines are monobasic only, for charge delocalization over both nitrogens, in the manner of an amidinium cation, effectively prevents addition of a second proton. The effect of the delocalization is strongest in 4-aminopyridine (pK_a 9.1) and much weaker in 2-aminopyridine (pK_a 7.2). Delocalization of this kind is not possible for the 3-isomer, which thus can form di-cations in excess concentrated acid (pK_{a_1} 6.6, pK_{a_2} −1.5).

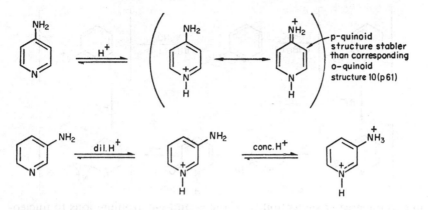

Whereas monoalkylation occurs mainly on the ring nitrogen, acetylation gives the product of reaction on the exocyclic nitrogen. The alkylation is

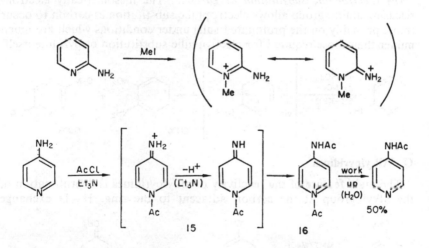

not reversible at room temperature, so the kinetically controlled product of reaction at the most basic nitrogen is isolated. In acetylation, however, it is the product of attack at the amino group which is isolated: after initial acylation at ring nitrogen, deprotonation of the NH_2 group by base (acetate or Et_3N in an $AcCl/Et_3N$ reagent) leads to a highly reactive enamide species, 15, which then acylates at the exocyclic nitrogen to give the diacetyl cation, 16. Hydrolysis of the ring acyl group during work up finally gives the observed product.

The only other noteworthy reaction in this group is with nitrous acid, which with β-aminopyridine gives a normal diazonium salt, but with α- and γ-aminopyridines yields the corresponding pyridones in dilute aqueous acid, and the bromopyridines in conc. HBr. This difference is

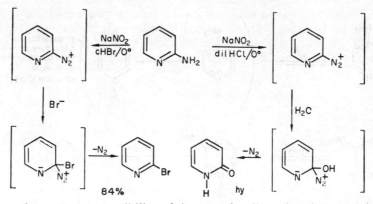

due to the greater susceptibility of the α- and γ-diazonium ions to nucleophilic attack at the α- and γ-carbons, and parallels the greater reactivity
of α- and γ-halopyridines when compared with the β-isomers.

(b) *Electrophilic substitution at carbon.* The mesomerically electron-
releasing amino group allows electrophilic substitution at carbon to occur
(most probably on the protonated salt) under conditions which are much
milder than those required for electrophilic substitution of pyridine itself.

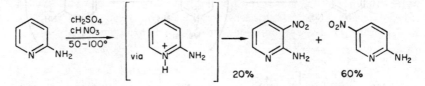

G   Alkylpyridines

The main feature of the reactivity of alkylpyridines is deprotonation of
the alkyl group at the carbon adjacent to the ring. H—D exchange

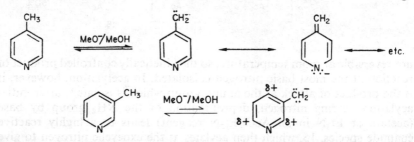

experiments in MeOD-MeO⁻ show an order of reactivity 4—>2—≫3—
in a ratio of 1810 : 130 : 1. The much greater ease of deprotonation of the
α and γ isomers is once again to be seen in the stabilization of the negative charge by mesomerism with the ring nitrogen, a process not available
to the β-isomer. In the latter, stabilization can only occur by induction by
the electron-deficient pyridine ring.

These anions react as nucleophiles in a wide range of reactions, and are closely analogous to enolate anions in reactivity. Similar activation of aralkyl groups is seen in *o*-nitrotoluene and to a lesser extent in *p*-nitrotoluene (but not in *m*-nitrotoluene).

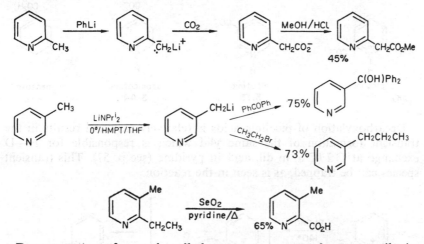

Deprotonation of α- and γ-alkyl groups occurs much more easily in pyridinium ions, when it leads reversibly to low equilibrium concentrations of nucleophilic enamine-like species. This is clearly seen in the $ZnCl_2$ or Ac_2O catalysed condensations with benzaldehyde.

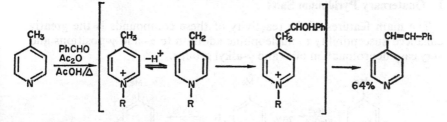

Reaction can even occur in very weakly basic conditions.

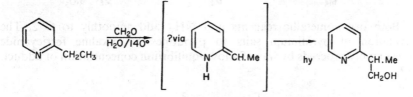

H Pyridine Aldehydes, Ketones, and Carboxylic Acids

These compounds all closely resemble corresponding benzene compounds largely because a carbonyl group cannot interact mesomerically with the pyridine nitrogen.

The three pyridine carboxylic acids exist almost entirely in their zwitterionic forms in aqueous solutions, and are slightly stronger acids than benzoic acid.

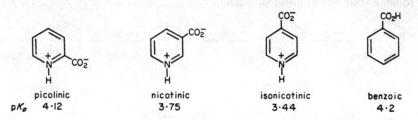

	picolinic	nicotinic	isonicotinic	benzoic
pK_a	4·12	3·75	3·44	4·2

Decarboxylation of picolinic acids is relatively easy and results in the transient formation of the same ylid which is responsible for H—D exchange at C 2 (C 6) in dil. acid in pyridine (see p. 57). This transient species can be trapped, as is seen in the reaction.

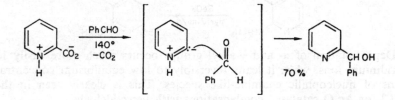

I Quaternary Pyridinium Salts

The main features of the reactivity of these compounds is the greatly enhanced susceptibility to nucleophilic addition to α- and γ-positions and very easy deprotonation of α- and γ-alkyl groups.

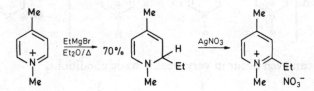

Both organometallic reagents and NH_2^- add smoothly to C 2. The oxidation of pyridinium salts to pyridones by alkaline ferricyanide presumably proceeds by way of a low equilibrium concentration of adduct, 17.

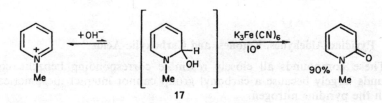

17

Borohydride reduces by initial transfer of hydride to C 2, followed by further reduction to yield mainly 1,2,5,6-tetrahydropyridines (3-piperideines), as does Na/EtOH reduction of pyridines (see p. 59), with some fully reduced piperidine: the latter can also arise by initial transfer of

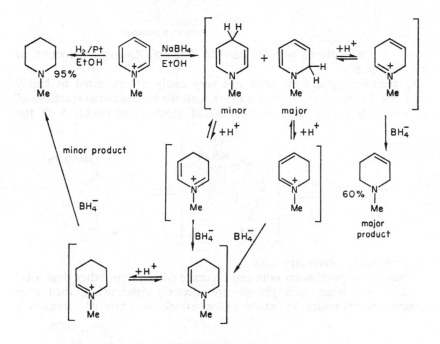

hydride to the C 4 of the pyridinium cation. The reduction in simple cases cannot be stopped at the dihydro stage: however, if an electron-withdrawing group is present at C 3 this stabilizes a 1,4-dihydropyridine, as seen in the example. Such reactions are being extensively studied because

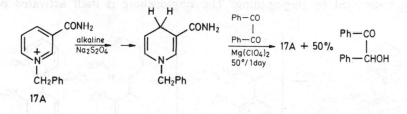

of the very important role played by nicotinamide derivatives in reducing and oxidizing enzymes. Model reductions have now been achieved in the laboratory; the reduction of benzil is one such example. Recent studies suggest that these redox reactions proceed by one-electron transfer steps.

A very special case of reduction is seen in the zinc reduction of 1-ethyl-4-carbomethoxy pyridinium cation to a stable mesomeric free radical

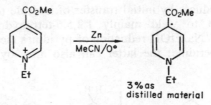

which can actually be isolated and distilled (cf. Zn-Ac$_2$O reduction of pyridine, p. 60).

Quaternary pyridinium salts are very easily deprotonated to highly nucleophilic enamines, which means that all the condensation reactions of α- and γ-alkyl pyridines can be effected much more readily with the

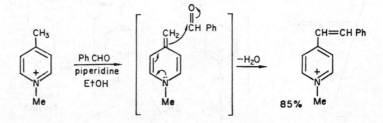

corresponding quaternary salts.

Quaternary pyridinium salts can be made with groups other than alkyl on nitrogen: when such groups are electron-withdrawing, then even greater susceptibility to nucleophilic attack is observed, frequently

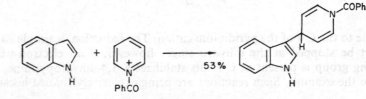

accompanied by ring-opening. The ring-opening is itself activated by

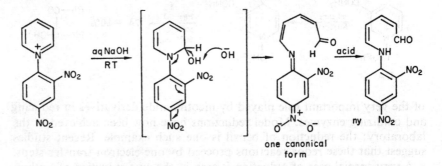

the electron withdrawal of the N-substituent.

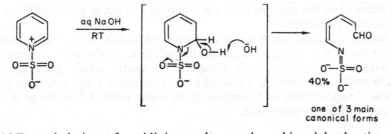

one of 3 main
canonical forms

N-Demethylation of pyridinium salts can be achieved by heating in dimethyl formamide solution with or without triphenylphosphine.

J   Pyridine N-Oxides

Pyridine N-oxides are of great interest partly because of the several ways in which their reactions differ from those of simple pyridines and partly because they have been found to be useful as intermediates in the synthesis of pyridines.

The most striking difference between pyridines and the N-oxides is the much greater susceptibility of the latter to electrophilic nitration. This is believed to be due to mesomeric electron release by the oxygen, which is formally analogous with electron release by phenoxide oxygen. This view is strongly supported by a comparison of the dipole moments of the following compounds: the difference of 2·03 D between pyridine and its N-oxide

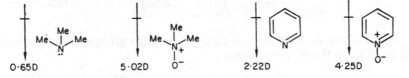

is much smaller than that of 4·37 D between trimethylamine and its N-oxide. This smaller difference suggests that canonical forms 18 and 19

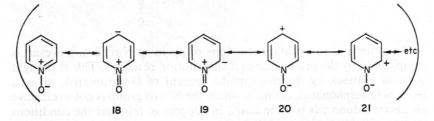

make significant contributions in the mesomeric system. The complexity of the situation, however, is evident when considering canonical structures 20 and 21 which must also contribute to the resonance hybrid: these canonical structures are an indication of the versatility of the polarizability of the molecule, for it can be seen that the N-oxide grouping can act both as an electron acceptor and as an electron donor capable of facilitating, on demand by the reagent, both electrophilic and nucleophilic addition to the α- and γ-positions.

As examples one can mention nitration, which has been closely studied and shown to proceed by electrophilic addition to the neutral pyridine N-oxide, and reaction with Grignard reagents, which proceeds by nucleophilic addition to C 2 followed by electrocyclic ring opening to give oximes; recyclization and elimination of water result when the oxime is heated with acetic anhydride.

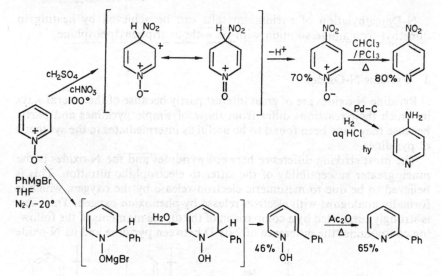

Nucleophilic displacement of halogen or nitro groups occurs at least as easily as in the pyridine series.

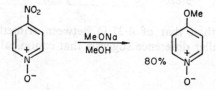

The pattern of substitution reactions of pyridine N-oxides, however, is complicated by the occurrence of β substitution reactions. This is thought to involve attack by the electrophilic reagent of O-protonated, or even perhaps O-sulphonated, cation in which the N⁺OR group is not as effective an electron donor. It is to be noted in support of this that the conditions necessary to bring about sulphonation are comparable to those necessary

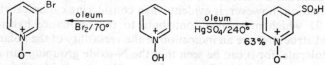

to sulphonate pyridine itself.

Deprotonation of α- and γ-alkyl groups occurs about as readily as with quaternary pyridinium salts and in consequence condensation reactions

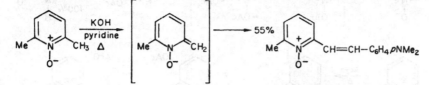

occur easily, even with the relatively unreactive p-dimethylamino-benzaldehyde.

Pyridine N-oxides are very weak bases, with pK_a values around 1, and of course protonate on oxygen. Alkylation also occurs on oxygen, and the resulting N-alkoxypyridinium salts are very susceptible to nucleophilic addition, as is shown in the following example:

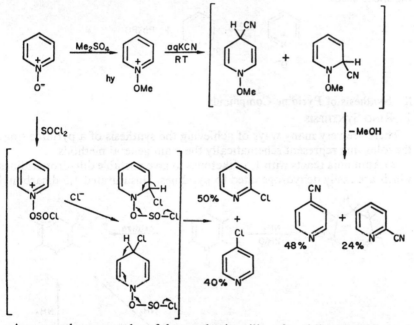

Among other examples of the synthetic utility of pyridine N-oxides one may cite reaction with thionyl chloride or phosphorus oxychloride which provides an easy route to 2- and 4-chloropyridines, and reaction with acetic anhydride which can lead to 2-acetoxypyridines, easily hydrolysed to give 2(1H)-pyridones in high yield. 3-Substituted pyridine-N-oxides usually give mixtures in which oxygen is introduced at the two alternative α-positions: 3-picoline-N-oxide for example gives a 1:1 mixture of 3-methyl-2- and 6-(1H)-pyridones. Acetic anhydride treatment of α- and γ-alkyl-pyridine N-oxides results in the preparatively useful acetoxylation of the side chain, although β-acetoxylation competes with this process, particularly in the γ-alkylpyridine N-oxides.

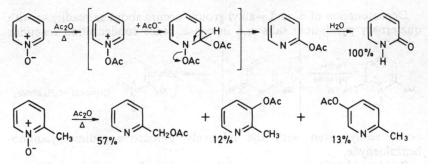

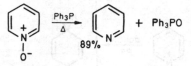

Several methods are available for the removal of N-oxide oxygen. Processes which employ transfer of oxygen to a trivalent phosphorus compound, such as triphenylphosphine or triethyl phosphite, have been the most used.

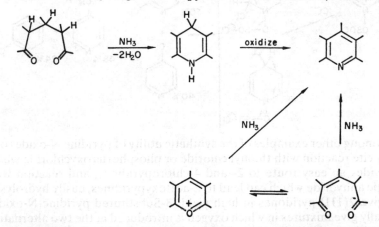

K　Synthesis of Pyridine Compounds

1　RING SYNTHESIS

There are very many ways of achieving the synthesis of a pyridine ring: the following represent schematically the main general methods.

(*a*) Ammonia reacts with 1,5-diketones to give unstable dihydropyridines which are easily dehydrogenated to pyridines. Unsaturated 1,5-dicarbonyl

compounds lead directly to a pyridine as do pyrylium salts. The usefulness i limited only by the availability of an appropriate dicarbonyl compound

(*b*) Symmetrical pyridines are often simply prepared by the inter action of ammonia, a β-dicarbonyl compound, and an aldehyde, followec by dehydrogenation.

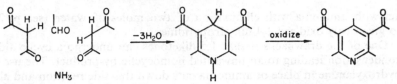

(c) Unsymmetrical pyridines result from the quite different interaction

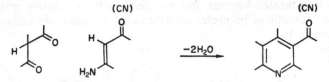

of a β-dicarbonyl compound and enaminone or nitrile.

(d) A very similar reaction using cyanoacetamide leading to 3-cyano-2-pyridones, easily transformed into pyridines, has found wide application.

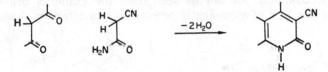

(e) A Diels-Alder reaction between an oxazole and a suitable dienophile is a more recent and very versatile method.

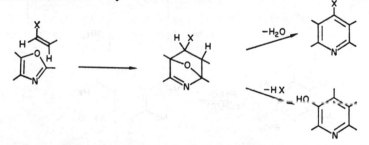

(a) *From 1,5-dicarbonyl compounds.* 1,5-Diketones are fairly readily available by a number of routes, such as Michael addition, as in the typical synthesis shown, or by ozonolysis of a cyclopentene precursor. They inter-

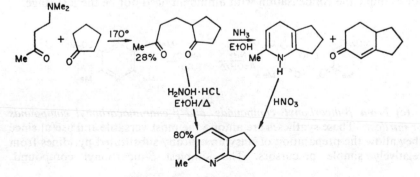

act with ammonia with elimination of two moles of water (see p. 27) to give easily oxidized 1,4-dihydropyridines.

One of the drawbacks is that 1,5-diketones can undergo a cyclic aldol condensation leading to an unwanted homocyclic by-product. The use of hydroxylamine in place of ammonia cuts down this side reaction and also eliminates the oxidation step.

Since unsaturated ketones and pyrylium salts are closely related, the latter can sometimes be useful intermediates, if easily synthesized themselves (see p. 160).

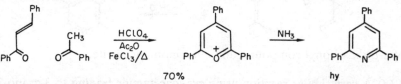

70%   hy

(b) *From aldehyde, two moles of β-dicarbonyl compound and ammonia. Hantzsch Synthesis.* As can be seen from the examples below, the product of this type of approach is necessarily symmetrically substituted

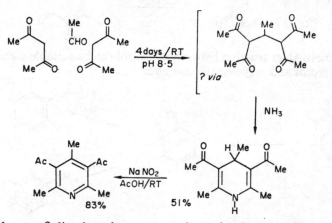

83%   51%

since the one β-dicarbonyl component is used twice, the aldehyde carbon becoming the pyridine γ-carbon.

Here again, the precise sequence of the intermediate steps is not known, for example the condensation with ammonia need not be the last stage.

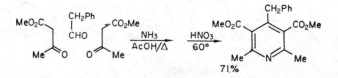

71%

(c) *From β-dicarbonyl compounds and β-enaminocarbonyl compounds or nitriles.* These syntheses are among the most versatile and useful since they allow the preparation of unsymmetrically substituted pyridines from relatively simple precursors. The simplest β-dicarbonyl compound, malondialdehyde, is too unstable to be used, but its readily available acetal enol ether can be used instead, as the following example shows.

24%

The enaminoketones are very easily prepared by the reaction of ammonia with the corresponding β-dicarbonyl compound, pentan-2,4-dione in the example above.

In the cases where the β-dicarbonyl component is unsymmetrical, two isomeric pyridines can result. However, the two carbonyl groups may have a sufficiently different reactivity to direct the ring synthesis entirely the one

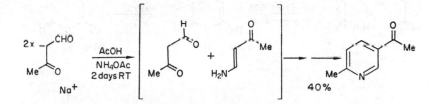
40%

way. Thus, if one mixes the sodium salt of formyl acetone with ammonium acetate in acetic acid, only one product is formed. The more reactive aldehyde carbonyl group reacts with the ammonia to give specifically one enamine, and moreover this enamine then reacts in a specific manner with a second molecule of the dicarbonyl compound.

Note that great care must be taken not to confuse these reagents and reactions with those employed in the Knorr synthesis of pyrroles (see p. 213).

Of the very many variants in this group of syntheses, the following two are instructive. Both involve conjugate addition of the enamine com-

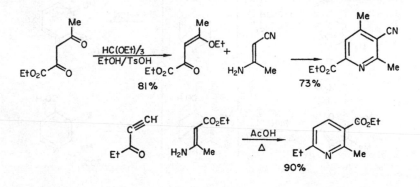

ponent. In the first case the prior formation of an enol ether makes it possible effectively to reverse the orientation in which this unit combines (compare with example below).

(d) From β-dicarbonyl compounds and cyanoacetamide

Guareschi Synthesis. This approach is probably the most widely used. It has found many important applications, notably one of the early commercial vitamin B_6 syntheses (see p. 80). The reactants are a β-dicarbonyl compound again, and cyanoacetamide. These react together under mildly basic conditions to give good yields of 3-cyano-2-pyridones. The nitrile and oxygen functions are valuable in that they are open to easy elaboration into a variety of substituent groups or can be eliminated as shown in the example. Notice once again that the differential reactivity of the two carbonyl groups results in the formation of only one of the two possible isomeric products.

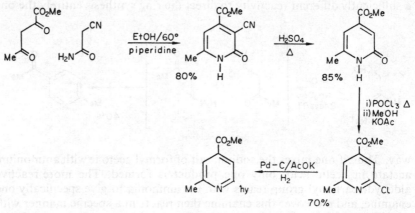

(e) From oxazoles and dienophiles.

An important recent development involves the cycloaddition of an oxazole, as the diene component, to a dienophile to give an intermediate which can then be aromatized either by loss of water or, if the dienophile is maleonitrile or acrylonitrile, by loss

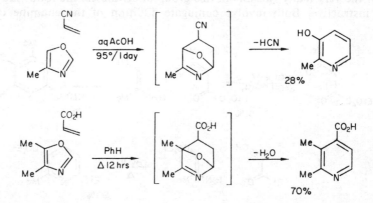

of hydrogen cyanide. This approach has led to successful syntheses of vitamin B_6.

(*f*) *Miscellaneous methods of ring synthesis.* Many alkylpyridines are manufactured commercially by chemically complex processes which often produce them together in mixtures. A classical example is the Chichibabin synthesis in which acetaldehyde is heated with ammonium acetate.

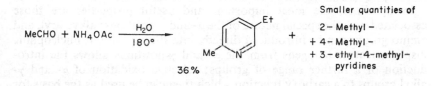

Mechanistically the process is probably related to the syntheses in group (*c*).

No account of pyridine synthesis would be complete without mention of the many ways in which furans can be exploited, by way of oxidation to 2,5-dioxy-2,5-dihydro derivatives. As a typical example we have the preparation of 6-propyl-3-hydroxypyridine.

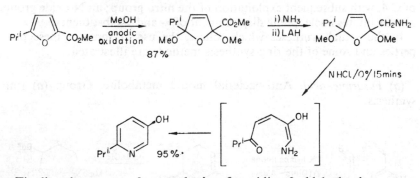

Finally, the spectacular synthesis of pyridine-3-aldehyde deserves a special mention as an example of the many unclassifiable routes to simple

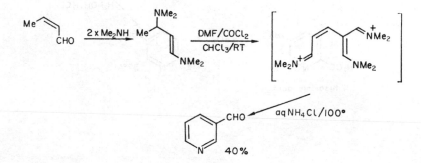

pyridines. Effectively it is a diformylation of crotonaldehyde with the Vilsmeier reagent, ending up with a group (*a*) cyclization.

2   FURTHER ELABORATION OF PYRIDINE COMPOUNDS

As was discussed in the general introduction to synthesis, once an appropriate intermediate is either bought or made by ring-synthesis, the reactivity of the ring or of the substituents then has to be exploited in order to complete the preparation of the desired compound.

In pyridines, the most important and useful properties are those associated with the special nature of the α- and γ-positions: alkyl, aryl, and amino groups can be introduced directly into the α-position; nucleophilic displacement of halogens from the α- and γ-positions allows the introduction of a further range of groups: selective oxidation of α- and γ-alkyl groups to a carboxy function which then can be used as the basis for the building-up of a side-chain by standard methods; finally, α- and γ-alkyl groups can also be used to build up side-chains by exploiting the nucleophilic reactivity of the corresponding mesomeric carbanions. More akin to benzene chemistry is the use of amino and oxy substituents to activate ring positions to electrophilic substitution, the activating group can then generally be replaced by hydrogen.

N-Oxide formation is also synthetically very useful in allowing nitration of C 4, with subsequent exploitation of the nitro group; an N-oxide group also facilitates nucleophilic displacement of α- and γ-substituents.

In the actual syntheses which follow, the use of some of these properties and some of the ring-synthesis methods are illustrated.

(*a*) *Fusarinic acid.* Anti-bacterial mould metabolite. Group (*a*) ring synthesis.

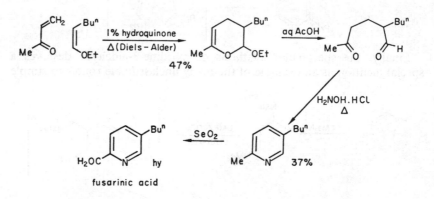

fusarinic acid

(*b*) *Pyridoxine.*   Vitamin B_6. Group (*d*) ring synthesis.

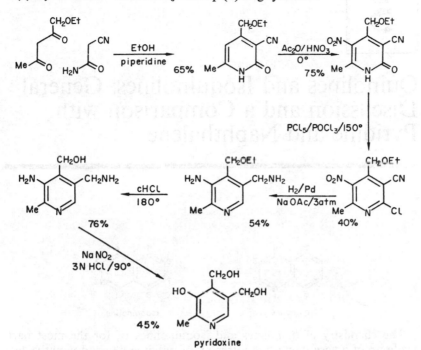

(*c*) *2-Methoxy-4-methyl-5-nitropyridine.*   An intermediate for synthesis of porphobilinogen (see p. 215).

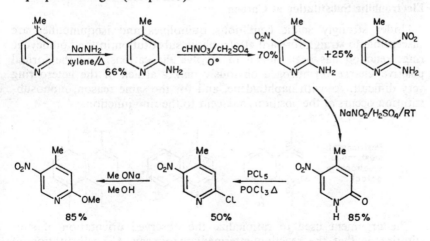

5

Quinolines and Isoquinolines: General Discussion and a Comparison with Pyridine and Naphthalene

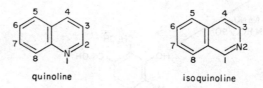

quinoline   isoquinoline

The chemistry of quinolines and isoquinolines is, for the most part, made up of a very interesting interplay of pyridine-like and naphthalene-like behaviour, and its analysis brings out many very useful lessons.

Electrophilic Substitution at Carbon

Under strongly acidic conditions quinolines and isoquinolines are susceptible to straightforward electrophilic substitution in the homocyclic ring: reaction has been shown to involve the cations, and the formal positive charge on nitrogen obviously makes attack of the hetero ring very difficult. As with naphthalene, and for the same reason, monosubstitution occurs at the positions adjacent to the ring-junction.

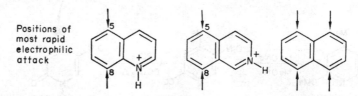

Positions of most rapid electrophilic attack

The argument used to rationalize the observed orientation of substitution is that the reaction intermediate of, say, C 5 substitution of quinoline, 1, has two favoured canonical forms with an unperturbed pyridinium system, with the second positive charge at C 6 in one and at C 8 in the other; on the other hand the reaction intermediate of, say, C 6 substitution, 2, has one such canonical form, with the second positive charge on C5 only; all other canonical forms contain quinoid structures and contribute much less to the hybrid structure: it is, then, the greater delocalization of the second positive charge in the homocyclic ring in the intermediate of C 5 substitution that accounts for its greater stability.

Electrophilic substitution in quinolinium and isoquinolinium occurs very much more readily than in pyridinium, but very much more slowly than in naphthalene; in fact the rates are comparable with those of tri-methylanilinium.

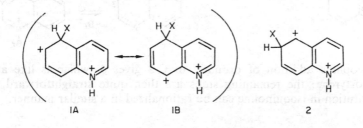

Partial rate factors for nitration

naphthalene (C 2)	benzene	quinoline (C 5, C 8)	isoquinoline (C 5)	trimethyl-- anilinium (C 3)
2×10^5	1	4×10^{-7}	9×10^{-6}	2×10^{-8}

(A partial rate factor is the ratio of the rate of substitution at a particular position in an aromatic system and the rate at one of the positions in benzene itself.

In the case of quinoline the partial rate factor for substitution at, say, C 5 would be given by:

$$\left(\frac{k \text{ (observed) for quinoline} \times \% \text{ attack at C 5}}{k \text{ (observed) for benzene} \times 1/6} \right).$$

The overall picture is greatly complicated by the ready occurrence of electrophilic substitution in the hetero ring at C 3 in quinoline and at C 4 in isoquinoline in weakly acidic conditions. This apparently glaring anomaly finds an explanation in the much greater ease with which nucleophilic addition occurs to the bicyclic systems when compared with pyridine (this is discussed under nucleophilic substitution): thus, taking bromination by Br_2 in CCl_4, the reaction is believed to proceed as shown in the following sequence:

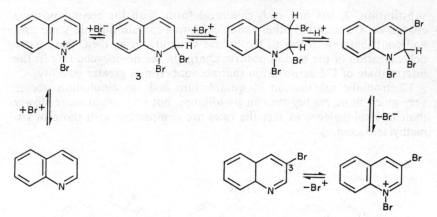

Nucleophilic addition of bromide to C 2 gives 3, which is like an *o*-aminostyrene; the remaining steps are then quite straightforward. C 4 substitution in isoquinoline can be rationalized in a similar manner.

Nucleophilic Substitution

Nucleophilic substitution of hydrogen predictably occurs at C 2 and, to a lesser extent, at C 4 in quinoline, and at C 1 in isoquinoline; C 3, the other 'α-position' in isoquinoline, is unreactive—this anomaly will be discussed later.

The most notable aspect of the reactivity of quinolines and isoquinolines towards nucleophiles is that it is much greater than that of pyridines: thus both quinoline and isoquinoline readily add NH_2^- in liquid ammonia (see pp. 92 and 111) whereas pyridine under these conditions does not react at all; the difference is further seen in the conditions given below for the Chichibabin aminations. It is also evident from the observation that

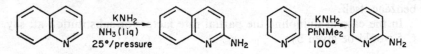

2-chloroquinoline reacts with sodium ethoxide about 300 times faster than does 2-chloropyridine.

This high level of reactivity can be explained as follows: the three main canonical forms for the three possible bicyclic intermediate anions, shown below, all contain an unperturbed benzene ring; given that the

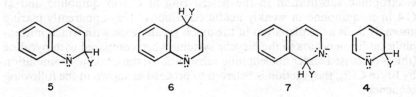

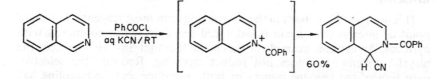

aromatic resonance energies of quinoline and isoquinoline are likely to be appreciably less than the sum of the aromatic resonance energies of benzene and pyridine, it follows that the loss in aromatic resonance energy in going from pyridine to the intermediate anion, 4, will be more than the corresponding loss in going from quinoline to 5 or 6, or from isoquinoline to 7.

This argument also applies to the greater ease with which quinolinium and isoquinolinium cations form *addition* products, whether with nucleophilic reagents, as in the Reissert reactions, or by what may be a pericyclic concerted reaction in the following reversible addition of vinylethyl ether to 2,3-dimethylisoquinolinium cation.

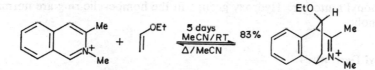

It is important to note that a parallel argument is used to rationalize the greater ease of electrophilic substitution in naphthalene when compared with benzene:

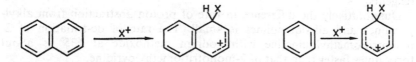

The very low reactivity of C 3 in isoquinoline towards nucleophilic reagents can now be understood to be the consequence of the fact that the canonical form, 8, in which the negative charge of the intermediate anion is on the nitrogen, does not contain a benzenoid ring: this raises the energy

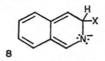

8

of this intermediate anion, and hence raises the activation energy of its formation.

Reduction

This occurs preferentially in the hetero ring with most reagents. The main point of interest once again is enhanced reactivity when compared with pyridine. An extreme example of this is the fact that Sn—HCl reduces the bicyclic systems, but does not reduce pyridine. Recently the selective reduction of the benzene moiety in both quinoline and isoquinoline has been achieved by the use of Pt/H_2 in strongly acidic solution.

Oxy- and Amino-derivatives

The pattern of tautomerism for groups in the hetero ring is closely similar to that found in pyridine, with the one exception of 3-hydroxyisoquinoline

of which the hydroxy tautomer is of the same order of stability as the carbonyl tautomer. Hydroxy groups in the homocyclic ring are normally phenolic.

Alkyl Derivatives

As in pyridine, alkyl groups α and γ to the ring-nitrogen are much more acidic, and undergo H-D exchange and the usual condensation reactions. Here again, although α to ring-nitrogen, alkyls at C 3 in isoquinolines are not appreciably activated, the reasons for which run parallel to those already given to account for the low reactivity of C 3 to nucleophilic reagents.

Quantitatively the difference in rate of proton abstraction from alkyl-pyridines and alkylquinolines is small. The rate of de-tritiation of 2-monotritiomethylquinoline with sodium methoxide at 135° is about forty times faster than that of 2-monotritiomethylpyridine.

Proton is easily lost from quaternary salts, and the neutral deprotonated products from quinolinium and isoquinolinium salts, in contrast with those from pyridinium salts, are stable enough to be isolated.

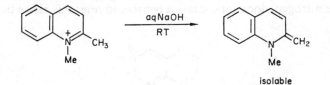

isolable

Quinolines: Reactions and Synthesis

Quinoline is a stable high-boiling liquid with a sweetish odour. It finds rather limited use in synthetic chemistry as a basic solvent, especially in Cu-catalysed decarboxylations.

It was first isolated from coal tar bases in 1834, and a little later in the alkaline pyrolysis of cinchonamine, an alkaloid closely related to the famous antimalarial alkaloid quinine. The word quinoline in fact is derived from the word quinine, which in turn is derived from *quina*, a Spanish version of a local South American name for the bark of quinine-containing *Cinchona* species. The subsequent importance of quinoline is linked with malaria in the several successful synthetic antimalarial drugs such as chloroquine, which is also used in the treatment of amoebic dysentery.

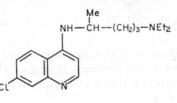

chloroquine

Quinolines play no part in fundamental metabolism, and they occur relatively rarely in plants as secondary metabolites (alkaloids), quinine

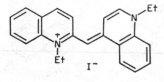

ethyl red

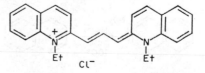

pinacyanol

being much the best known. An important role played by quinoline compounds was that of providing the first photographic film sensitizers, such as the cyanine dye 'ethyl red' which extended photography from the blue into the green and then in 1904, with pinacyanol, into the red.

Since that time, hundreds of sensitizing dyes have been made and investigated, and the quinoline nucleus has been pushed aside by other, more efficient, systems.

REACTIONS AND SYNTHESIS OF QUINOLINES

A Reactions with Electrophilic Reagents

1 ADDITION TO NITROGEN

All the reactions noted in this category for pyridine (p. 45) occur with quinoline and little further comment is necessary. With a pK_a of 4·94, it is of the same order of basicity as pyridine and isoquinoline.

2 SUBSTITUTION AT CARBON

(a) *Proton exchange.* The relative rates of H-exchange at the various quinoline positions are dependent on the acidity of the medium and the activity of the water in it. In 40 per cent aqueous sulphuric acid exchange is observed only as 240° is approached, and it then occurs mainly at C 2 and C 3: very surprisingly, hydroxyl anion is believed to play the main role by acting in two ways with 1-H-quinolinium cation, by C 2 deprotonation to generate the ylid 1 (cf. pyridines, p. 57) and by C 4 addition followed by the steps shown, which are quite similar to those leading to C 3 bromina-

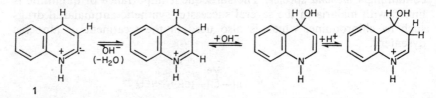

1

tion. As the acidity increases, with concomitant decrease in water activity, the simple protonation of the homocyclic ring of 1H-quinolinium cation becomes the dominant process, and in 90 per cent sulphuric acid it occurs at 180°, principally at C 8.

(b) *Nitration* occurs easily; it produces a high yield of a mixture of 5- and 8-nitroquinolines with no detectable quantity of any other isomers.

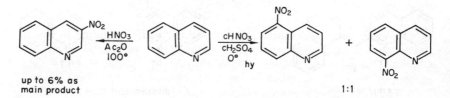

up to 6% as
main product   1:1

Kinetic data are consistent with electrophilic attack by the nitrating species on the quinolinium cation.

In acetic anhydride nitration proceeds in a totally different manner, and very inefficiently—much quinoline is recovered and much is converted into intractable products. The main product obtained under these conditions is the 3-isomer (1·9–6·6 per cent) together with a small quantity of a mixture of 6- and 8-nitroquinolines (0·7–0·9 per cent). A very significant point is that no 5-nitroquinoline is produced. All this seems to be in keeping with electrophilic substitution of a 1,2-adduct, for in such a molecule electron-release from nitrogen would be expected at C 3, C 6, and C 8.

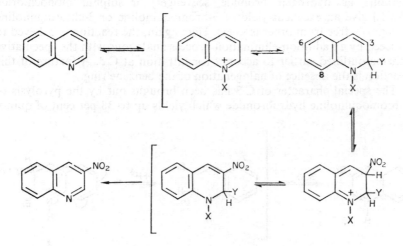

(c) *Sulphonation* gives the 8-sulphonic acid with only small quantities of the 5-isomer. This contrasts with nitration and halogenation but parallels mercuration (p. 90).

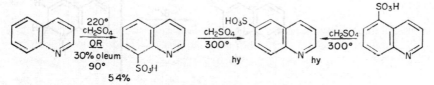

With a very high reaction temperature the 6-sulphonic acid is produced, but then under these conditions the 5- and 8-acids rearrange to the 6-acid and so it seems likely that such a rearrangement provides the pathway to this isomeric product.

(d) *Halogenation.* Bromination has been carefully studied in recent years, and a most interesting and complicated picture is emerging. Chlorination has received less attention, but from what is known the pattern of behaviour seems to be similar to that of bromination.

(i) The most straightforward reaction occurs by the addition of bromine

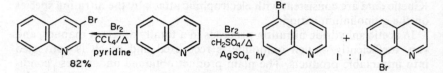

to excess quinoline in conc. sulphuric acid-silver sulphate, which leads to a good yield of an equimolar mixture of 5- and 8-bromoquinolines. This parallels nitration in strong acid and represents simple bromonium ion (or solvated bromonium ion) attack on quinolinium cation.

(ii) Bromine in carbon tetrachloride solution (best with the presence of pyridine as hydrogen bromide scavenger) or sulphur monochloride (S_2Cl_2) give an excellent yield of 3-bromoquinoline or 3-chloroquinoline practically free from other isomers. Here again, the reaction is believed to proceed by an addition-elimination process analogous with the speculative scheme outlined earlier to account for nitration at C 3. A feature of this reaction is the absence of halogenation of the benzene ring.

The special character of C 3 has been brought out by the pyrolysis of 3-bromoquinoline hydrobromide which yields up to 38 per cent of quino-

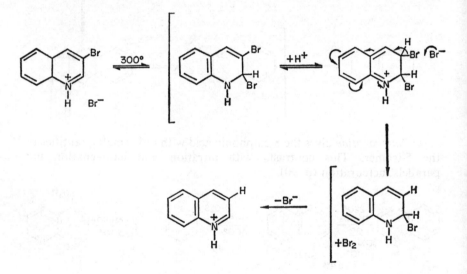

line (together with some bromine and a complex mixture of polybromo-compounds). This extraordinary dehalogenation, which does not occur with 6-bromoquinoline, is thought to proceed as shown. It is immediately obvious that such a scheme represents a reversal of the proposed bromination process for the formation of 3-bromoquinoline.

(e) *Mercuration.* Quinoline with mercuric acetate forms an N-mercuriacetate salt at room temperature, which rearranges at 160° to a mixture from which, on treatment with sodium chloride, the 8- and 3-chloromercuriquinolines can be isolated in unspecified yield. When the

8-position is blocked by a methyl group then the main product is 5-chloromercuri-8-methylquinoline.

(f) Acylation and alkylation. There seem to be no reports of successful Friedel-Crafts reactions on quinoline.

B   Reactions with Oxidizing Agents

Quinoline is resistant to most mild oxidizing agents, even aqueous chromic acid reacts only slowly. Potassium permanganate, however, attacks the homocyclic ring to give pyridine-2,3-dicarboxylic acid as

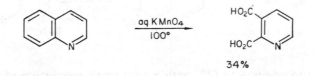

34%

main product. The alkyl groups of substituted quinolines can be oxidized to the corresponding quinoline carboxylic acid.

Calcium hypochlorite oxidizes quinoline to 2-quinolone. This could occur by an initial addition of chlorine to N, then as in the following scheme.

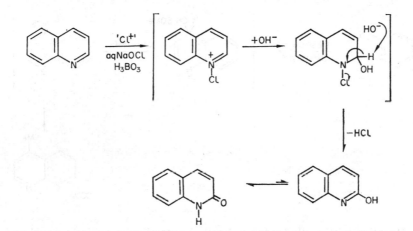

Peracids give rise smoothly to the N-oxides in a manner exactly analogous to their reaction with pyridines.

C   Reactions with Nucleophilic Reagents

1   SUBSTITUTION WITH HYDRIDE TRANSFER

(a) Alkylation and arylation. The simplest and best-studied reactions are those with alkyl- and aryl-lithium compounds and they lead almost

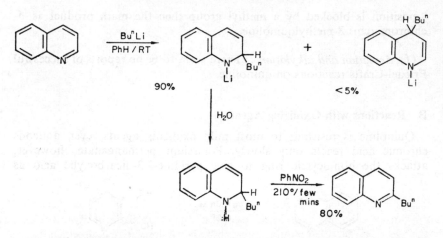

exclusively to addition to C 2, with a very small proportion of addition to C 4. The N-lithio-1,2-dihydroquinolines are hydrolysed to the corresponding bases which are reasonably stable compounds. Effective overall substitution and the formation of 2-substituted quinolines is then achieved only by subsequent oxidation, best by a mild hydride acceptor such as nitrobenzene.

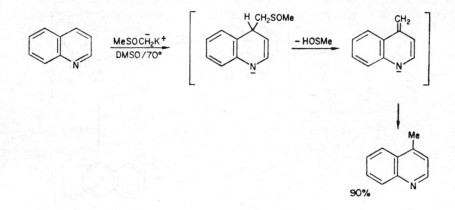

Reaction with methyl sulphinylmethyl potassium provides a novel and high yield alkylation at C 4. Pyridine does not react under these conditions.

(b) *Amination.* Quinoline reacts rapidly with $NaNH_2$ in liquid NH_3 to give a 3:1 mixture of α- and γ-adducts. At RT, reaction goes further to give 2-aminoquinoline: the nature of the cation influences the yield, which is highest (80 per cent) when $Ba(NH_2)_2$ is used. That 4-aminoquinoline is not formed in this reaction suggests that the reversibly formed γ-adduct suffers hydride loss much less readily under these conditions;

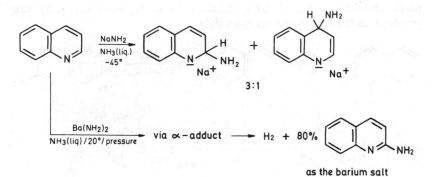

however, up to 10 per cent 4-aminoquinoline may be obtained in the presence of NaNO₃ as hydride acceptor. When substitution cannot occur at C 2, as with 2-phenylquinoline, the 4-amino derivative is produced.

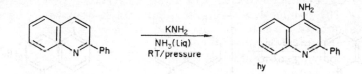

(c) *Hydroxylation.* When quinoline is heated with KOH or NaOH—KOH, 2-quinolone is produced together with a nearly-quantitative yield of hydrogen.

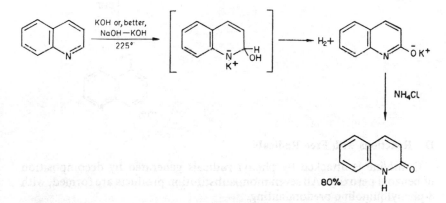

As with amination, the cation is of importance, for reaction with sodium hydroxide occurs only at about 300° and is much less efficient.

2   SUBSTITUTION WITH DISPLACEMENT OF HALIDE

Quinolines with halogen at C 3, C 5, C 6, C 7 and C 8 have normal halobenzene type reactivity, that is, resistance to substitution by nucleophilic reagents. 2- And 4-haloquinolines, however, react easily by substitution, as the following examples show.

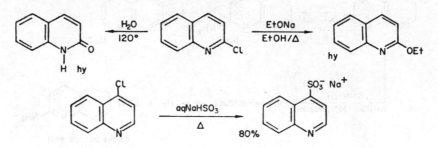

None of the fourteen bromo- and chloroquinolines appears to have been made to react with magnesium or lithium; however, lithio derivatives can be obtained from both homo- and hetero-ring bromides by reaction with two moles of butyllithium at a temperature low enough ($-70°$) to suppress addition.

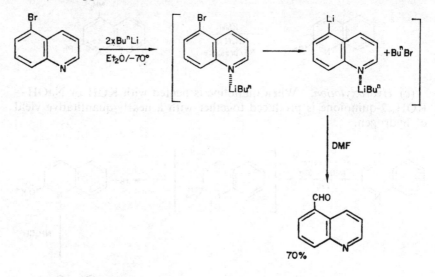

D   Reactions with Free Radicals

Quinoline is attacked by phenyl radicals generated by decomposition of benzoyl peroxide. All seven monosubstitution products are formed, with 8-phenylquinoline predominating.

substitution position	2	3	4	5	6	7	8
% of product	6	14	20	12	8	8	30

A much greater degree of specificity occurs when the quinolinium cation is attacked, and by less reactive radicals. Carboxamide and acetyl radicals have been introduced into the quinoline 2- and 4-positions in this way.

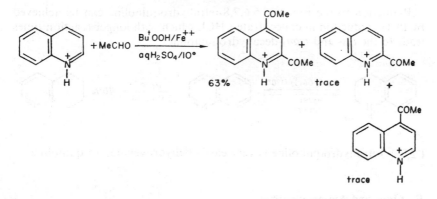

E   Reactions with Reducing Agents

Most reducing agents attack the hetero-ring, and many of them give high yields of 1,2-dihydroquinoline, for example lithium aluminium hydride (80 per cent), and diethylaluminium hydride (95 per cent).

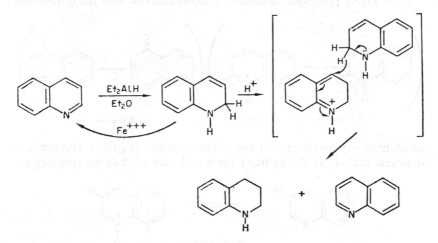

1,2-Dihydroquinoline is easily oxidized back to quinoline, best by ferric ion, but even by air. In acidic solution and in the absence of air, it disproportionates smoothly to a mixture of tetrahydroquinoline and quinoline. Reduction to 1,4-dihydroquinoline occurs with sodium or lithium in liquid NH_3: the unique aspect of this reduction product is that it is a simple enamine which might have been expected to equilibrate with the tautomeric 3,4-dihydroquinoline, which one would have anticipated to be the thermodynamically stabler tautomer.

Reduction to 1,2,3,4-tetrahydroquinoline is best achieved by catalytic hydrogenation with Raney nickel or palladium, though this result is also obtained by reduction with tin-hydrochloric acid and sodium-ethanol.

Reduction to the isomeric 5,6,7,8-tetrahydroquinoline can be achieved by Pt/H$_2$ reduction in concentrated HCl, which with longer reaction times leads to the *cis-* and *trans*-decahydroquinolines.

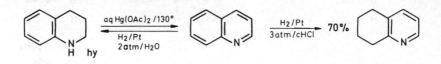

1,2,3,4-Tetrahydroquinoline is very easily dehydrogenated to quinoline.

F   Oxy- and Aminoquinolines

1   QUINOLINOLS AND QUINOLONES

Oxy-quinolines carrying the oxygen at C 2 or C 4 follow the pattern which emerged for pyridine: they exist, for all practical purposes, entirely in the carbonyl form.

These highly polarized mesomeric systems contrast with the quinolinols

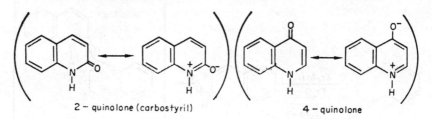

2 – quinolone (carbostyril)   4 – quinolone

which have oxygen situated at one of the other five positions. Oxygen here is almost entirely in the hydroxy form and may exist to varying degrees

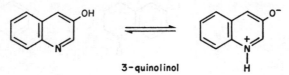

3 – quinolinol

in tautomeric equilibrium with the corresponding zwitterionic structure, as shown for 3-quinolinol.

The quinolinols are true phenols, they are *not* highly polarized, and when the hydroxyl is on the benzene ring they show all the reactions of the corresponding naphthols.

8-Hydroxyquinoline is the best-known compound in this group and has long been used in analysis as a chelating agent, especially for Zn^{II}, Mg^{II} and Al^{III}; the Cu chelate is a useful fungicide.

The 2- and 4-quinolones react with aqueous sodium hydroxide by N-deprotonation, and with concentrated strong acids by O-protonation, as shown for 2-quinolone.

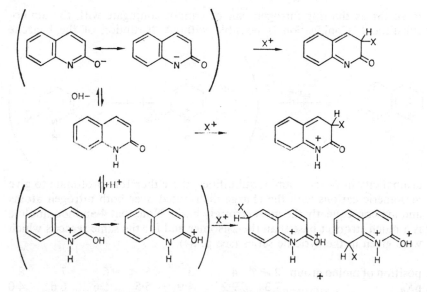

The favoured position for electrophilic substitution of neutral 4-quinolones and of the mesomeric anions is C 3, whereas in the protonated species attack occurs in the homocyclic ring at C 6 and C 8.

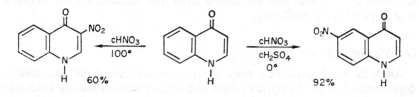

By contrast 2-quinolone, as the neutral, unprotonated molecule, reacts at C 6 preferentially. H-D exchange in strong aqueous sulphuric acid

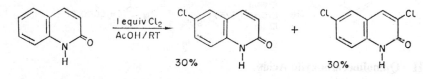

comes out slightly differently in favouring C 8 over C 6: the order is C 8 > C 6 > C 5 ~ C 3.

As with the pyridones, only oxygen at C 2 or C 4 can be replaced by halogen with phosphorus halides. The quinolinols do not react with this reagent type in this way.

2 AMINOQUINOLINES

As one might expect, all the monoaminoquinolines exist as the amino-tautomers, protonate on the ring nitrogen first, and differ in basicity only in so far as the ring nitrogen can or cannot conjugate with the amino-substituent. Conjugation is possible without disruption of the benzene

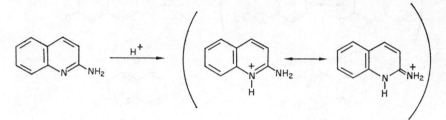

aromaticity in 2- and 4-aminoquinolines, these therefore protonate to give mesomeric cations with the charge distributed over both nitrogen atoms and are consequently the most strongly basic. Note that 4-aminoquinoline is a much stronger base than the 2-isomer and for the same reasons which were given in the pyridine series (see p. 64).

position of amino group	2	4	3	5	6	7	8
pK_a	7·3	9·2	4·9	5·5	5·6	6·6	4·0

All the amines are alkylated by methyl iodide on the ring nitrogen and are acetylated on the NH_2. The 5-, 6-, 7-, and 8-aminoquinolines are normally diazotizable and are brominated and nitrated on the benzene ring at predictable positions.

G Alkylquinolines

As with alkyl pyridines, only 2- and 4-alkylquinolines are readily deprotonated at the benzylic carbon and undergo condensation reactions as shown.

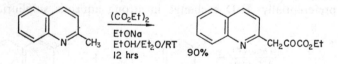

H Quinolinecarboxylic Acids

Only the 2-carboxylic acid readily decarboxylates, thermally at 160 °C: the reaction occurs by way of a transient ylid (cf. p. 67).

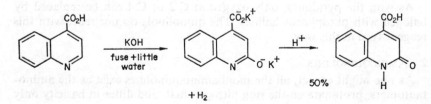

Worth noting is the stability of the carboxyl group at the quinoline 4-position, which survives, for example, even potassium hydroxide fusion.

I   Quinolinium Salts

By far the most important aspect of the reactivity of salts of this type is the greatly enhanced susceptibility to addition of a nucleophile to C 2 and in some instances to C 4.

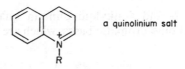

a quinolinium salt

For the simple proton salts (R=H) N-deprotonation by a nucleophile generally competes successfully with addition to carbon, however, we have seen (p. 90) what an important and somewhat complex role nucleophilic addition to quinolinium ions plays in electrophilic substitution at C 3.

The simple alkyl quinolinium salts (R=alkyl) undergo addition reactions easily, even with hydroxide ion. The actual product obtained depends on various structural factors; this first adduct may be the main product or

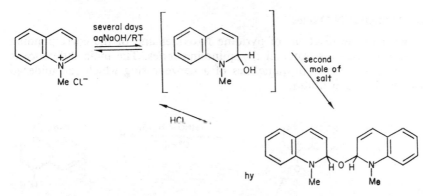

may undergo further reaction as is seen in the case of N-methylquinolinium cation. The dimeric carbinolamine ether is almost certainly produced by addition of the carbinolamine O-anion to a second quinolinium cation.

The specific addition of cyanide anion, not to C 2 but to the alternative

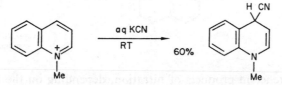

C 4, is an excellent example of the complexity of additions to ambient systems.

The Reissert reaction complicates matters still further by showing that in N-acyl quinolinium salts, cyanide addition occurs mainly to C 2.

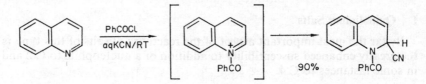

Deprotonation of 2- and 4-alkylquinolinium quaternary salts yields isolable very reactive homo-aromatic bases, which are highly susceptible to

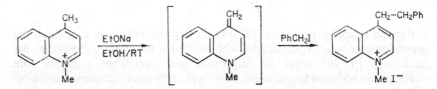

electrophilic addition on the exocyclic carbon and behave like enamines. They are the reactive intermediates which lead to the synthesis of the cyanine dyes (see p. 87).

J Quinoline N-Oxides

Most of the reactions of pyridine N-oxides and of the corresponding O-substituted salts occur in the quinoline series. The main difference is due to the presence in quinolines of a benzene ring which can undergo electrophilic substitution.

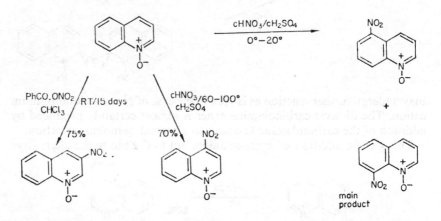

The difference in products of nitration, depending on the temperature of reaction, gives some indication of what subtle factors must be involved in determining the position of substitution.

The many reactions of O-alkyl N-oxides with nucleophilic reagents here, as in pyridine chemistry, open up a range of synthetically useful routes: cyanide can be introduced at C 2.

K   Synthesis of Quinoline Compounds

1   RING SYNTHESIS

There are three important methods for the construction of the quinoline ring system from non-heterocyclic precursors, and all three start with benzene compounds.

(*a*) Anilines react with 1,3-dicarbonyl compounds to give intermediates which can be cyclized with acid.

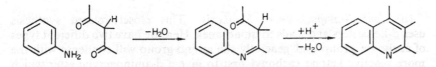

(*b*) Anilines react with an α,β-unsaturated carbonyl compound in the presence of an oxidizing agent to give quinolines. Most often glycerol is

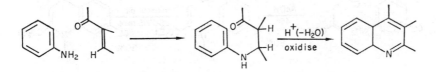

used as an *in situ* source of acrolein, and this method is the best for preparing quinolines carrying no substituents on the heterocyclic ring.

Because of the electrophilic character of the closure step in approaches (*a*) and (*b*), anilines carrying donating substituents *meta* to the amino group give rise to 7-substituted products. In approach (*b*) electron withdrawing groups situated *meta* to the amino group give 5-substituted quinolines.

(*c*) *ortho*-acyl anilines react with ketones to give quinolines directly.

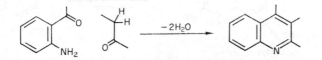

(*a*) *From arylamines and 1,3-dicarbonyl compounds.*

(i) *The Combes Synthesis.* Condensation of a 1,3-dicarbonyl compound with an arylamine gives a high yield of the β-amino-enone, which can then be cyclized with concentrated acid. The cyclization step is an electrophilic substitution by the mesomeric O-protonated amino-enone followed by loss of water to give the aromatic quinoline.

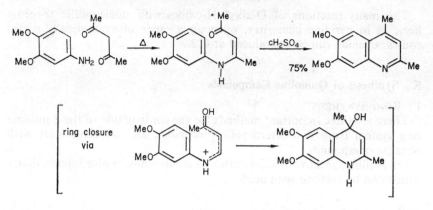

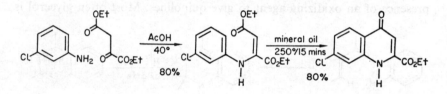

(ii) *Conrad-Limpach-Knorr Synthesis.* This closely-related synthesis uses β-ketoesters and leads to quinolones. Here we have two different types of carbonyl group and generally the amino group will condense with the more reactive ketone carbonyl first to give a β-aminoacrylic ester which then can be thermally cyclized at 250° to give a 4-quinolone.

Although the arylamine condenses with the ketonic carbonyl group at low temperature (kinetic control), this reaction is reversible, and at higher temperatures the stabler amide (thermodynamic control) is formed after interaction of the arylamine with the ester group. This second type of condensation product can be cyclized to an isomeric 2-quinolone, and thus one can control the orientation of cyclization.

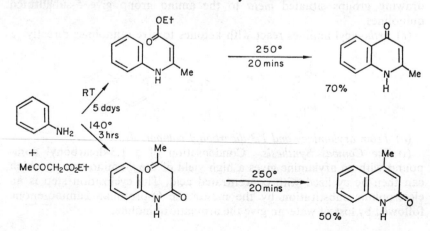

(*b*) *From arylamines and α,β-unsaturated carbonyl compounds.*

The Skraup Synthesis. In this extraordinary reaction, quinoline is produced when aniline, concentrated sulphuric acid, glycerol and a mild oxidizing agent are heated together. The reaction has been shown to proceed by dehydration of the glycerol to acrolein to which aniline then adds β. Acid-catalysed cyclization then gives 1,2-dihydroquinoline. Nitrobenzene or arsenic acid are generally used as oxidizing agents.

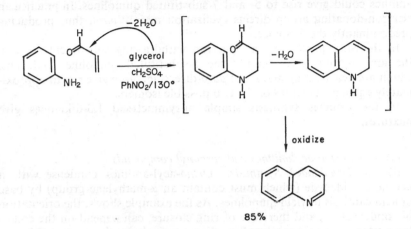

The use of substituted carbonyl components, as in the following example, shows that the intermediate is not formed by condensation of the amino group with the carbonyl.

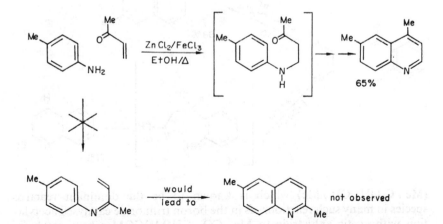

Skraup reactions sometimes become too vigorous, and care must be taken to control the temperature. The method cannot be used on compounds which contain acid-labile substituents. Apart from this limitation, it is the best method available for the synthesis of quinolines unsubstituted on the hetero-ring.

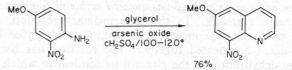

76%

Orientation of Ring Closure in Approaches (a) and (b). *Meta*-substituted anilines could give rise to 5- and 7-substituted quinolines. In practice an electron-donating group directs cyclization mainly *para*, thus producing predominantly the 7-isomer.

In the Skraup reaction, an electron withdrawing substituent *meta* to the amino-group gives mainly the 5-substituted quinoline. Halogens, which are deactivating and *ortho/para*-directing, often give rise to approximately equal proportions of the two possible isomers.

In the Combes synthesis simple unsymmetrical 1,3-diketones give mixtures.

(c) From ortho-acylanilines and carbonyl compounds.
(i) *The Friedländer Synthesis.* *Ortho*-acyl-anilines condense with a ketone or aldehyde (which must contain an α-methylene group) by base or acid catalysis to yield quinolines. As the example shows, the orientation of condensation, and therefore of ring closure, can depend on the conditions used. The acid-catalysed reaction proceeds by the neutral C 3 enol

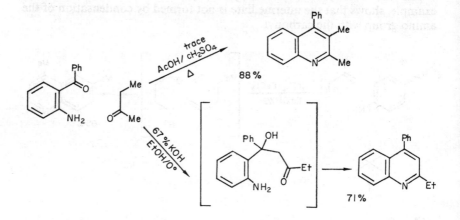

(Me . C(OH):CH . Me) which is known to be the dominant reactive species in many such reactions, as in the boron trifluoride catalysed acetylation with acetic anhydride to Me . CO . CH(Me)COMe; base catalysis, on the other hand, generates the C 1 enolate anion, Me . CH₂ . CO . CH₂⁻ as in the well-known iodoform reaction.

(ii) *The Pfitzinger Synthesis.* The requisite *ortho*-aminobenzaldehydes are often not easy to obtain, and this modification of the Friedländer approach uses substituted isatins, which are relatively easily synthesized.

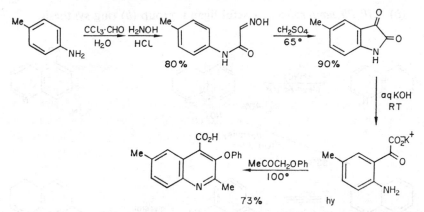

Alkali opens isatins to *ortho*-aminophenylglyoxylate anions, condensation of which with appropriate ketones then gives quinoline-4-carboxylic acids.

Many such compounds have been prepared for transformation into derivatives for pharmacological testing because of their structural similarity to quinine. If not required, the 4-carboxyl group can be eliminated by pyrolysis with calcium oxide.

2   QUINOLINE SYNTHESES WITH ELABORATION OF SUBSTITUENTS

The following examples give some idea of the sort of manipulations which, based on a knowledge of quinoline chemistry, are used in the syntheses of active quinoline compounds.

(*a*) *Chloroquine.* Synthetic antimalarial, for ring synthesis (see p. 102).

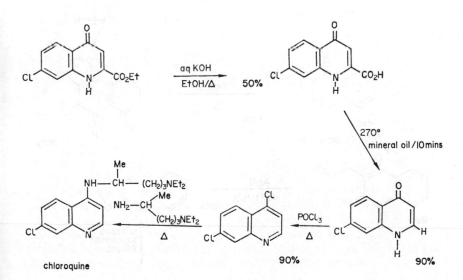

(b) *1,10-Phenanthroline.*   Useful ligand, group (b) ring synthesis.

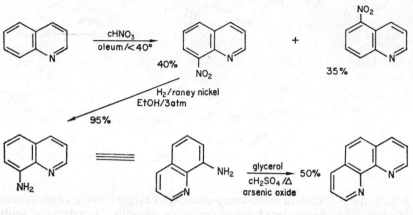

(c) *γ-Fagarine.*   Alkaloid, group (a) ring synthesis.

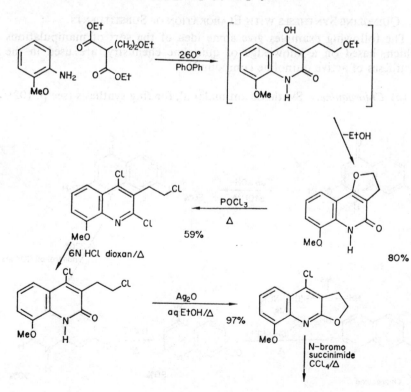

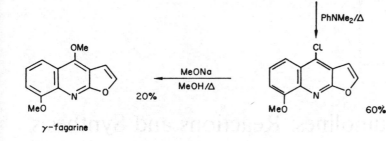

γ-fagarine

7

Isoquinolines: Reactions and Synthesis

Isoquinoline is a stable, low-melting solid with a sweetish odour. It was first isolated in 1885 from the quinoline fraction of coal-tar bases.

Neither it nor its simple derivatives are used in organic chemistry in any general way. It is not involved in fundamental metabolism, nor is it incorporated in any synthetic chemotherapeutic agent. This rather negative record is somewhat offset by the opium poppy alkaloid papaverine, which is a dehydrogenated variant of the very large group of secondary plant metabolites known as the isoquinoline alkaloids nearly all of which are 1,2,3,4-tetrahydroisoquinoline derivatives. Papaverine is one of the most potent coronary vasodilators. Morphine and emetine, both tetrahydroisoquinoline derivatives, are examples of medicinally-valuable members of this group of alkaloids.

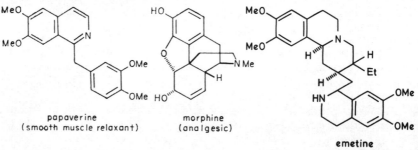

papaverine
(smooth muscle relaxant)

morphine
(analgesic)

emetine
(amoebicide)

REACTIONS AND SYNTHESIS OF ISOQUINOLINES

A   Reactions with Electrophilic Reagents

1   ADDITION TO NITROGEN

In reactions involving donation of the nitrogen lone pair to electrophiles, isoquinoline is entirely comparable with pyridine and quinoline. As with quinoline, the pK_a of isoquinoline (5·4) is similar to that of pyridine.

2   SUBSTITUTION AT CARBON

(a) *Proton exchange.*   Proton exchange in isoquinoline closely parallels that of quinoline (see p. 88) and occurs in the 2H-isoquinolinium cation at C 1 by deprotonation, and at C 4 by C 1 addition of OH^- in 40 per cent sulphuric acid at 240°. At higher acidity, in 90 per cent sulphuric acid at 180°, it occurs by simple protonation of the 2H-isoquinolinium cation, more rapidly at C 5 than at C 8.

(b) *Nitration* occurs smoothly to give mainly the 5-substituted derivative. Nitration involves electrophilic attack on isoquinolinium cation. It is

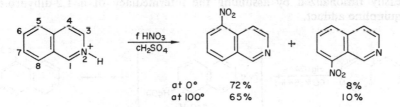

at 0°   72%   8%
at 100°   65%   10%

noteworthy that the isoquinolinium cation is nitrated about ten times more rapidly than is quinolinium cation.

The nitration of 1-benzylisoquinoline gives 1-(4′-nitrobenzyl)-isoquinoline, nicely illustrating the inherently deactivated nature of the isoquinoline homoaromatic ring compared with the isolated benzene nucleus.

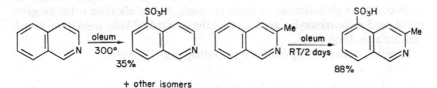

35%   88%

+ other isomers

(c) *Sulphonation* occurs readily and almost exclusively at C 5. At high temperatures substitution is much less specific. With oleum at 300° isoquinoline gives only a moderate yield of the 5-sulphonic acid, the rest being a mixture containing mainly the 8-sulphonic acid.

The formation of the 5-sulphonic acid has been shown, like nitration, to involve attack of the isoquinolinium cation.

(d) *Halogenation.*   (i) Bromination occurs, depending on the conditions of reaction, at C 4 or at C 5. The substitution at C 4 may well be analogous

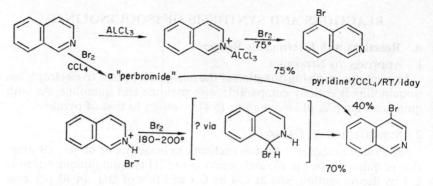

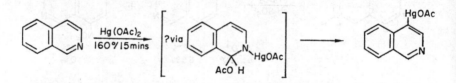

with substitution by halogen at C 3 in quinoline, and involve β-bromination of the enamine system of a 1,2-adduct.

(ii) Chlorination in the presence of aluminium chloride gives a slow conversion into 5-chloroisoquinoline.

(e) *Mercuration.* This occurs mainly at C 4 and once again is most easily rationalized by assuming the intermediacy of a 1,2-dihydroiso-quinoline adduct.

B   Reactions with Oxidizing Agents

Isoquinoline is largely resistant to the action of most oxidizing agents and is attacked notably by potassium permanganate. This, in neutral solution, gives phthalimide as main product, and in alkaline solution gives pyridine-3,4-dicarboxylic acid and phthalic acid. Thus, oxidative attack can occur in both rings.

Peracids convert isoquinolines smoothly into N-oxides.

C   Reactions with Nucleophilic Reagents

1   SUBSTITUTION WITH HYDRIDE TRANSFER

Reactions of this type occur exclusively at C 1. Nucleophilic substitution of hydrogen at C 3 is not known.

(a) *Alkylation and arylation.* Alkyllithium compounds add under mild conditions which allow survival of the 1,2-dihydroquinoline product, though, if required, subsequent conversion into the 1-substituted isoquino-line is very easy. Allyl Grignard reagents also react readily, but alkyl

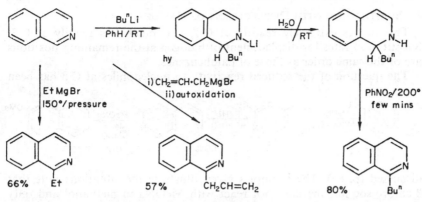

Grignard species require elevated temperatures to give moderate yields of aromatic substituted isoquinoline product.

Addition still occurs at C 1 even with isoquinolines already carrying a group at that carbon.

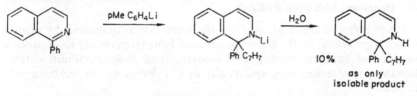

Methylation with methyl sulphinylmethylpotassium takes place at C 1.

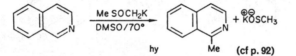

(b) *Amination.* Sodamide in liquid ammonia at −10° adds to C 1 to give the salt 1, which in time or at higher temperature loses hydride to give 1-aminoisoquinoline. There is no report of 1-phenylisoquinoline reacting with sodamide, which could indicate the non-reactivity of C 3.

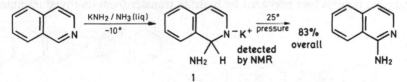

(c) *Hydroxylation.* Isoquinoline reacts with potassium hydroxide at high temperature and 1-isoquinolone is formed in good yield.

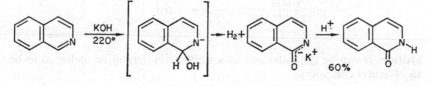

2   SUBSTITUTION WITH DISPLACEMENT OF HALIDE

This occurs extremely readily at C 1, and much less readily at C 3. Substitution rates for displacement of halogen at the remaining positions are of the same order as those of halobenzenes.

The question of the reduced reactivity to nucleophiles at C 3 has been

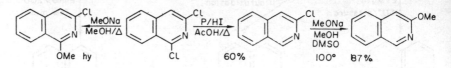

discussed (p. 85). The examples given illustrate the situation: note that 3-chloroisoquinoline does not react with MeONa in methanol and only does so in dimethylsulphoxide, a solvent well known greatly to facilitate ionic reactions. The formation of isoquinoline Grignards and lithium derivatives from haloisoquinolines has not been reported.

D   Reactions with Free Radicals

Isoquinoline is attacked, at about the same rate as quinoline, by phenyl radicals generated by the decomposition of benzoyl peroxide to produce a mixture of isomeric monophenyl-isoquinolines. Isoquinolinium cation, however, is attacked very specifically at C 1 by acetyl or carboxamide radicals.

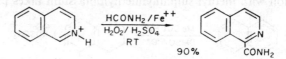

E   Reactions with Reducing Agents

Sodium and liquid ammonia or lithium aluminium hydride treatments of isoquinoline do not give clean reactions. 1,2-Dihydroisoquinoline, a very reactive base, is best prepared by hydride transfer from diethylaluminium

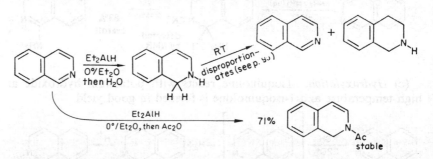

hydride. It can be characterized as a stable acyl-derivative, believed to be an N-acetyl compound.

Reduction with sodium in boiling alcohol gives 1,2,3,4-tetrahydro-isoquinoline; a slower reaction, but better yield, is obtained with tin-hydrochloric acid. Catalytic reduction (Pt) in acetic acid gives 1,2,3,4-tetrahydroisoquinoline but in concentrated HCl at 3 atm the product is the isomeric 5,6,7,8-tetrahydroisoquinoline; further slow reduction under these conditions leads to a 4:1 mixture of *cis*- and *trans*-decahydroiso-quinolines.

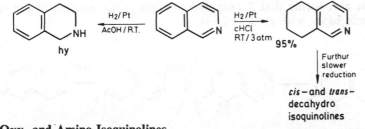

F   Oxy- and Amino-Isoquinolines

1   ISOQUINOLINOLS AND ISOQUINOLONES

The 4-, 5-, 6-, 7- and 8-isoquinolinols are true phenolic compounds, whereas 1-isoquinolone exists completely in the carbonyl form under all conditions.

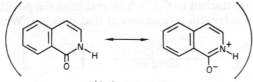

1(2H)− isoquinolone

The 3-isomer is of interest, because in this instance the two tautomers are of comparable stability. 3-Isoquinolinol is dominant in ether solution, 3-isoquinolone is dominant in aqueous solution. A colourless solution of

3−isoquinolinol   3−(2H)−isoquinolone
(colourless)   (yellow)

3-isoquinolinol in ether turns yellow on addition of a little methanol, because of the appearance of some of the carbonyl tautomer.

That the two tautomers do not differ greatly in stability is in large measure the consequence of the balancing of two opposing tendencies. The stable amide structure in 3-isoquinolone forces the benzene ring into an unstable quinoid structure, and the stable benzenoid ring in the 3-isoquinolinol necessarily means an unstable lactim structure in the hetero-ring. 1-Isoquinolone, on the other hand, has both benzenoid and amide units and change to the 1-isoquinolinol necessarily means a decrease in stability.

2 Aminoisoquinolines

Since after N-protonation none of the aminoisoquinolines can have polarized *para*-quinoid canonical structures *and* a benzene ring, the highest pK_a is that of 1-aminoisoquinoline, only 7·6 and close to that of 2-aminoquinoline at 7·3. However, the extended *para*-quinoid canonical structure of 6-aminoisoquinolinium cation gives it the highest pK_a of the benzene-substituted aminoisoquinolines and quinolines at 7·2.

It is noteworthy that 3-aminoisoquinoline is the weakest of the seven isomeric bases, with a pK_a of 5·0.

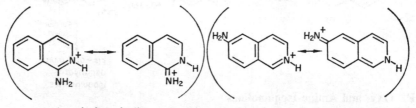

1–aminoisoquinoline ·protocation   6–aminoisoquinoline protocation

G Alkylisoquinolines

Here again, enhanced acidity of saturated C-hydrogen is observed when the alkyl group is attached to C 1, which results in the possibility of carrying out, easily, condensation reactions at that position. When attached to

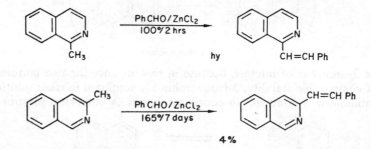

C 3 the acidity is very much decreased, but is still more marked than that at other positions.

H Isoquinolinecarboxylic Acids

Only the 1-carboxylic acid readily decarboxylates thermally. If the decarboxylation is carried out in the presence of benzaldehyde then the intermediate ylid can be trapped.

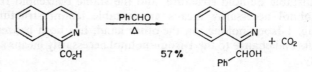

I Isoquinolinium Salts

Nucleophilic addition occurs only at C 1, and under mild conditions as the examples show.

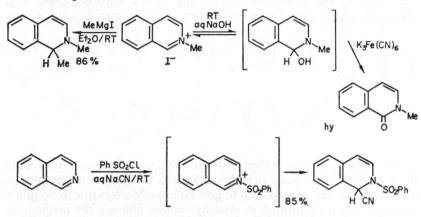

Alkyl groups at C 1 and C 3 in isoquinolinium cation are much more reactive than in the neutral isoquinoline system, and condensation reactions require correspondingly milder conditions.

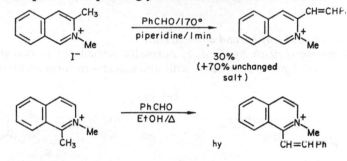

J Synthesis of Isoquinoline Compounds

1 RING SYNTHESIS

Most isoquinolines are synthesized by one of the following three general ring-forming routes.

(a) Benzaldehydes react with aminoacetal to give anils which can be cyclized with strong acid. This approach gives isoquinoline compounds

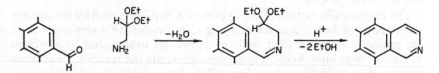

carrying no substituents on the hetero-ring. It can be used to obtain isoquinolines carrying electron-donating substituents at C 7 from *meta-*

substituted benzaldehydes, and is the best of the three methods for the preparation of isoquinolines carrying electron-withdrawing groups on the homocyclic ring.

(*b*) and (*c*) The other two general approaches involve the use of a phenylethylamine in condensation with an aldehyde or acid. A subsequent

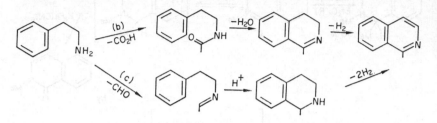

acid-catalysed ring closure then gives a di- or tetrahydroisoquinoline. Both these types can be readily dehydrogenated to the fully aromatic species. These methods can be employed to give exclusively 6-substituted isoquinolines from *meta*-substituted phenylethylamines, and are the methods of choice for the preparation of isoquinolines carrying alkyl and aryl groups at C 1. Neither of these two is efficient in producing isoquinolines from starting materials carrying electron-withdrawing groups on the benzene ring.

(*a*) *From benzaldehydes and aminoacetal.*

The Pomeranz-Fritsch Synthesis is normally carried out in two stages. First a benzaldehyde is condensed with aminoacetal to form an aldimine.

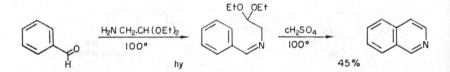

This step proceeds in high yield under mild conditions. Secondly the aldimine is cyclized by treatment with strong acid; hydrolysis of the imine competes and tends to reduce the efficiency of this step.

The last step in the sequence is of a familiar type (see Combes and Skraup syntheses) in that the acid initially causes the elimination of one mole of ethanol and the production of a species which can attack the ring as an electrophile. Final elimination of a second mole of alcohol completes the process.

The electrophilic nature of the cyclization step explains why the process works best for benzaldimines carrying electron-donating substituents on the aromatic ring and least well for molecules deactivated by electron-withdrawing groups. Activating groups accelerate the reaction more when *meta* to the aldehydic carbon than when alternatively placed. Their *meta* orientation allows mesomeric electron release to the site of electrophilic attack. In activating both *ortho* and *para* positions, the possible closure

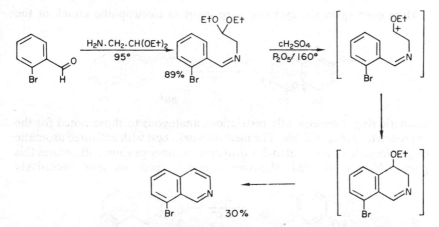

points, they accelerate the cyclization. The electrophilic carbonium ion has a choice of two positions for attack. Invariably ring closure takes place *para* to the activating group and 7-substituted isoquinolines are thus formed. This preference is thought to be associated with steric interference to *ortho* attack.

Isoquinolines substituted at C 1 are not easily formed by the Pomeranz-Fritsch procedure. The first step would require formation of a ketimine from aminoacetal and a ketone. This process does not proceed as well as reaction with a benzaldehyde. A variation of the approach which overcomes this difficulty has been developed. A substituted benzylamine is condensed with glyoxal diethyl acetal. The resultant imine, an isomer of that which would be produced by the conventional approach, can be cyclized in the normal way.

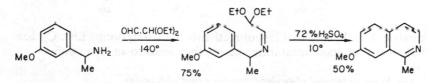

(b) *From phenylethylamides.*

The Bischler-Napieralski Synthesis.   A phenylethylamine reacts with a carboxylic acid or acid chloride to form an amide which can be cyclized, with loss of water, to a 3,4-dihydroisoquinoline. Common condensing

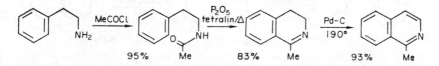

agents used include phosphorus pentoxide, phosphorus oxychloride and phosphorus pentachloride. Dihydroisoquinolines can be easily dehydrogenated to isoquinolines.

Here once again the cyclizing step involves electrophilic attack of the

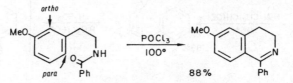

aromatic ring. Consequently restrictions analogous to those noted for the approach (*a*) are applicable. The method works best with activated aromatic rings; the yield of the 7-nitro-3,4-dihydroisoquinoline shown illustrates this point. *Meta*-substituted phenylethylamines react to give exclusively

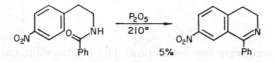

6-isomers, by attaching *para* and not *ortho* to the activating group. The method complements the approach (*a*) which is useful for the preparation of 7-substituted isoquinolines.

Pictet-Gams Modification. By carrying out the Bischler-Napieralski synthesis with a potentially unsaturated phenylethylamine, a fully aromatic heterocycle can be obtained directly. The amide of a β-methoxy- or β-hydroxy-β-phenylethylamine is heated with the usual type of condensation

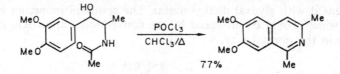

catalyst. It is thought that loss of methanol (or water) occurs first. Cyclization of the resultant unsaturated amide then leads to an isoquinoline.

(*c*) *From activated phenylethylamines and aldehydes.*

The Pictet-Spengler Synthesis. Phenylethylamines react with aldehydes easily and in good yields to give imines. Tetrahydroisoquinolines result from their cyclization since imines are in a lower oxidation state than the

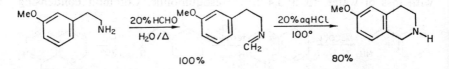

amides used as intermediates in the Bischler-Napieralski synthesis. Cyclization of an aldimine of phenylethylamine can, in favourable cases, be effected with acid. This cyclization is again electrophilic in character and is indeed a Mannich reaction. Protonation of the nitrogen atom produces a

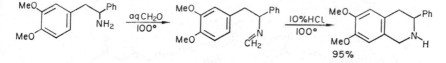

relatively weakly electrophilic immonium species. For efficient cyclization by this route then, powerful activating groups are necessary in the aromatic

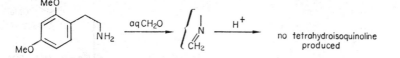

ring. A methoxyl group is not sufficient to cause reaction to proceed unless oriented *para* to the point of closure. The closing of the heterocyclic ring takes place in the same sense as it does in the Bischler-Napieralski approach, that is electrophilic attack takes place *para* and not *ortho* to the activating group.

Highly activated hydroxylated aromatic rings permit Pictet-Spengler ring closure under very mild, 'physiological' conditions. Such mild syntheses prompted early speculation on the biogenesis of tetrahydroiso-

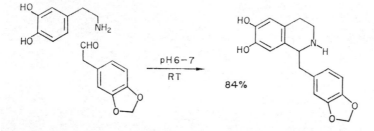

quinoline alkaloids. It was argued that the laboratory syntheses were utilizing the actual physiological conditions present within plant cells and were thus duplicating the biosynthesis of such alkaloids. It is now recognized that the biosynthesis of naturally-occurring compounds is enzyme-catalysed and less simple than was originally thought.

2   EXAMPLE OF ISOQUINOLINE SYNTHESIS

Papaverine, smooth muscle relaxant; useful as coronary vasodilator; alkaloid from opium. Type (*b*) synthesis, Pictet-Gams modification.

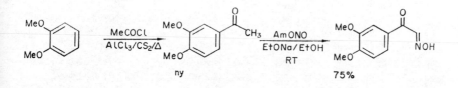

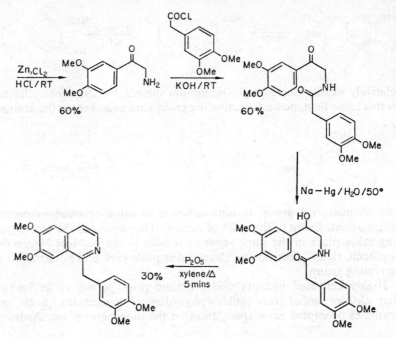

8

Quinolizinium Salts: General Discussion and a Comparison with Pyridinium Salts

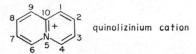

quinolizinium cation

It is of some interest briefly to consider the third possible azanaphthalene system, the quinolizinium cation. This system is completely isoelectronic with naphthalene, and the positive charge on the nitrogen is the consequence of its nuclear charge, which is one more than carbon.

In practically all its reactions it bears a close resemblance to a quaternary pyridinium salt: it is resistant to electrophilic substitution and reacts easily with nucleophilic reagents by addition to C 4 (α to the nitrogen) as the first step.

The reactivity of quinolizinium cation is such that it is not attacked by boiling water, but is attacked by piperidine, Grignard reagents, sodium borohydride, and LAH.

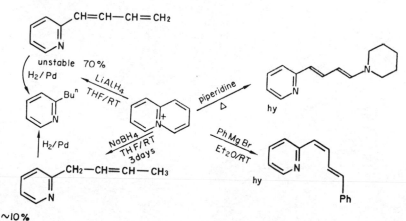

In all these reactions the initial adduct is not isolated, but ring-opens to a pyridine compound. This is, of course, reminiscent of the behaviour of

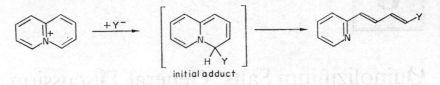

initial adduct

pyridinium salts with an electron-withdrawing group on the nitrogen (see p. 70) and of the adducts of pyrylium ions with nucleophiles (p. 149).

Further resemblance to pyridinium ions is to be seen in the acidity of methyl groups at C 2 and C 4, which may be condensed with, say, benzaldehyde to give styrylquinolizinium salts, and may be oxidized by selenium dioxide to carboxyl.

9

Diazines: General Discussion and a Comparison with Pyridines and s-Triazine

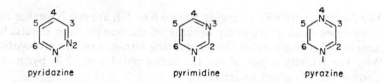

pyridazine   pyrimidine   pyrazine

The unsubstituted diazines are much more resistant to electrophilic substitution than is pyridine, thus they have not been nitrated or sulphonated. However, this generalization does not apparently apply to bromination of pyrimidine at C 5, for this occurs more readily than that of pyridine; bromination of pyrazine and pyridazine have not been reported. Alkyl substitution seems to exert a large influence: thus C 3 chlorination of 2-methylpyrazine occurs under surprisingly mild conditions. Whether we are dealing here with straightforward electrophilic substitution or with addition-elimination reactions is not known. Noteworthy in this connection is the observation that bromination of 4-phenylpyrimidine occurs exclusively in the pyrimidine ring: that no detectable bromination of the benzene ring occurs only really makes sense in terms of a mechanism involving bromination of a non-aromatic pyrimidine derivative, such as 1, a mechanism analogous with that of the C 3 bromination of quinoline (see p. 90). On the whole, our knowledge of the relative reactivities of the diazines and of their alkyl derivatives to electrophiles is still very fragmentary.

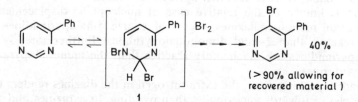

40%

(> 90% allowing for recovered material)

1

Available data suggest that the diazinones are more resistant to electrophilic substitution than the pyridones. 5-Hydroxy pyrimidine, the only phenolic diazine, has not been observed to undergo simple electrophilic substitution, probably because it is surprisingly acid labile, in contrast with 3-hydroxypyridine which is stable to acid and nitrates and sulphonates smoothly at C 2. That the bromination of 2(1H)-pyrimidone in aqueous acid proceeds by way of a non-aromatic intermediate bromo compound has been established by NMR: whether or not the bromination step involves the 2(1H)-pyrimidone itself remains to be proved.

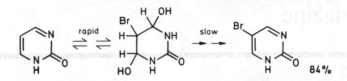

The diazines, with pK_a's ranging from 0·6 to 2·3, are much weaker bases than pyridine (5·2), presumably because of the destabilizing effects, both inductive and mesomeric, of the second ring nitrogen on the protocations.

Why the basicity drops along the series pyridazine, 2·3, pyrimidine, 1·3, pyrazine, 0·6, is not understood.

Survival of the diazines under oxidizing conditions is of the same order

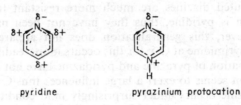

pyridine   pyrazinium protocation

as with pyridine and benzene, though no comparative experiments are reported.

The most important aspect of the chemistry of the diazines involves reactions with nucleophilic reagents. Here again, little is known of the more quantitative aspects of these reactions. Experimental reaction conditions and yields suggest that addition of organolithium compounds occurs at least as readily as with pyridine. Data on the amination of the unsubstituted diazines are practically non existent, but suggest pyrazine to be more reactive than pyridine. The greater reactivity of pyrimidine towards nucleophilic reagents is seen in its decomposition by concentrated aqueous alkali, whereas the other systems are stable.

More is known of the relative ease of nucleophilic displacement of halogen: all monohalodiazines, except of course 5-halopyrimidines, are more reactive than 2- and 4-monohalopyridines. Furthermore, 2- and 4-halopyrimidines are much more reactive than the monohalopyrazines and pyridazines.

Broadly speaking, then, the extra nitrogen in the diazines renders them more reactive towards nucleophiles than pyridine. In pyrazines and pyridazines the second nitrogen helps by generating electron deficiencies at atoms on which negative charge builds up during a nucleophilic reaction involving the first nitrogen. In pyrimidine, the second nitrogen assists much more strongly because it is as directly involved as the first in the substitution.

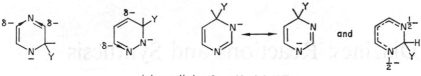

intermediates for attack by Y⁻

This is an appropriate point at which to mention symmetrical triazine. This heterocycle, though quite stable and aromatic in character, is so very susceptible to nucleophilic attack that it is rapidly decomposed by water: this is readily understood as an extrapolation of the reactivity of pyrimidine. Nucleophilic addition leads to a highly symmetrical intermediate in

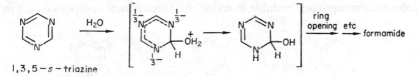

1,3,5 − s − triazine

which the negative charge is distributed over the three nitrogens, and thus greatly stabilized.

2,4,6-Trichloro-1,3,5-triazine, or cyanuric chloride, is a highly reactive halogen compound, almost like an acid chloride. It is industrially important in forming the basis of a wide range of complex dyestuffs (Procion and Cibacron dyes), not because of the chromophoric properties of the triazine nucleus, but because when substituted at C 2, C 4, and C 6 by N or O it forms a stable unit to which can be attached two or three dyestuffs, and which can even be covalently linked to fibre cellulose.

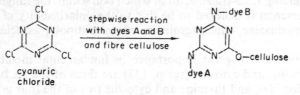

cyanuric chloride

stepwise reaction with dyes A and B and fibre cellulose

dye A

A family of herbicides, of which ametryne is a member, is based on the s-triazine nucleus; the very widely used melamine-formaldehyde resins are based on 2, 4, 6-triamino-1, 3, 5-triazine.

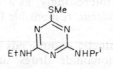

ametryne

Diazines: Reactions and Synthesis

The three diazines, pyrazine, pyrimidine, and pyridazine are stable colourless compounds, soluble in water. All three parent compounds are in

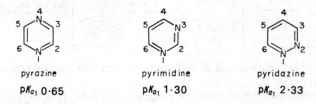

a way chemical curiosities, for none of them has found any application in general organic chemistry, nor is any used as a starting material for the synthesis of any of the useful derivatives. The one striking aspect of the physical properties of this group is the abnormally high boiling point of pyridazine which, at 207°, is 80–90° higher than any of the other simple azines, including 1,3,5-triazine, all of which boil within the range 114–124°: this phenomenon is believed to be due to the polarizability of the N—N system in pyridazine, which results in extensive dipolar association in the liquid.

Pyrimidines are of great importance in fundamental metabolism, for uracil, thymine, and cytosine (see p. 133) are three of the six bases found in the nucleotides, and thymine and cytosine two of the four crucial bases in nucleic acids (see p. 327); the pyrimidine ring also occurs in the vitamin thiamin (see p. 320). By contrast, no naturally-occurring pyridazine compound is known. The pyrazine system turns up in the fungal metabolite aspergillic acid, and, more interestingly, as a 1,4-dihydropyrazine in the luciferin of *Cypridina hilgendorfii*, responsible for the chemiluminescence of this ostracod.

Finally, many useful synthetic compounds contain a diazine system: for example, several sulphonamide drugs, such as sulphadiazine; the currently widely-used antimalarial pyrimethamine; and the selective plant

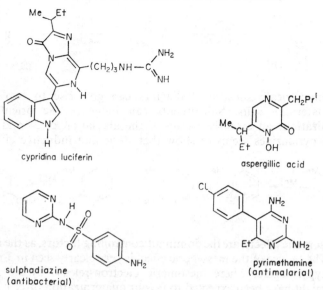

cypridina luciferin

aspergillic acid

sulphadiazine
(antibacterial)

pyrimethamine
(antimalarial)

growth regulator (used as a selective weed-killer on lawns) 3-hydroxy-6(1H)-pyridazinone (see p. 133), to name but a few.

REACTIONS AND SYNTHESIS OF DIAZINES

A   Reactions with Electrophilic Reagents

1   ADDITION TO NITROGEN

(a) *Protonation.* The diazines are essentially monobasic compounds and are, as is seen from the pK_{a1} values, much weaker than pyridine. This drop in basicity is believed to be largely the consequence of destabilization of the N 1-protonated cations by inductive election-withdrawal by the second nitrogen.

N,N′-diprotonation is very much more difficult and has been observed in very strongly acidic solution only with pyrazine ($pK_{a2} - 6·2$) and with pyrimidine ($pK_{a2} - 6·9$): the failure of pyridazine to protonate a second time on N 2 is most probably due to the high energy required to generate two positive charges on adjacent atoms.

Substituents such as methyl or methoxyl have base-strengthening effects comparable to those observed in pyridines, so that, for example, 4-methylpyridazine (pK_a 2·92) is a stronger base than pyridazine (pK_a 2·33); basicity drops quite sharply down the series 4-methoxy-6-methylpyrimidine (pK_a 3·65), 4-methylpyrimidine (pK_a 1·98), and pyrimidine (pK_a 1·30).

(b) *Alkylation.* The diazines react with alkyl halides easily to give monoquaternary salts. Dialkylation to diquaternary salts has not been achieved with alkyl halides; the very much more reactive trialkyloxonium borofluorides, however, do convert all three systems into the diquaternary salts.

86%   2 BF₄⁻   85%

Unsymmetrically-substituted diazines can give rise to two isomeric monoquaternary salts. Substituents can influence the orientation of quaternization: an extensive study of the alkylation of 3- and 3,6-substituted pyridazines seems to show that steric and inductive effects, and

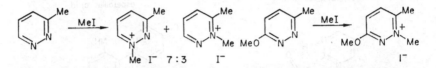

not mesomeric effects, are the dominant controlling factors, as the examples show. The failure of the mesomeric effect is very clearly seen in 3-methyl-6-methoxypyridazine, where mesomeric electron-release by the methoxyl group might have been expected to favour quaternization of N 1.

2 SUBSTITUTION AT CARBON

Remembering the resistance of pyridines to electrophilic substitution, it is not surprising to find that the introduction of a second azomethine nitrogen in any of the three possible positions greatly increases this resistance: thus so far no nitration or sulphonation of a diazine or of a simple alkyl diazine has been reported. However, halogenation is possible as is seen below in two of the very few known cases. Note that the C 5 of pyrimidine is the only position in all three diazines not to be in α or γ relation to a ring nitrogen.

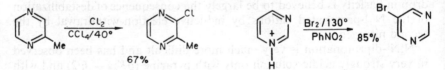

B Reactions with Oxidizing Agents

The diazines are generally resistant to oxidative attack at ring carbon, though pyrazine is oxidized by alkaline potassium permanganate at room temperature and is thus much more susceptible than pyridine. Good yields of diazine dicarboxylic acids are obtainable from appropriate benzoderivatives.

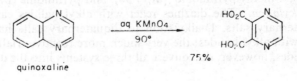

quinoxaline   75%

Hydrogen peroxide in acetic acid gives mono-N-oxides: predictably, di-N-oxide formation is most marked with pyrazines, but pyridazine gives only about 1 per cent.

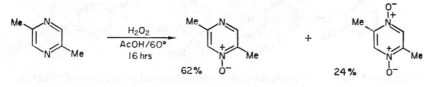

62%   +   24%

C   Reactions with Nucleophilic Reagents

The three diazines undergo H/D exchange at all ring positions with NaOMe–MeOD at 164°: this process, which occurs by deprotonation to transient carbanions, is somewhat faster than the corresponding one in pyridine (see p. 57).

The diazines are very susceptible to nucleophilic addition; for example, pyrimidine is destroyed when heated with aqueous alkali by a process which almost certainly involves nucleophilic addition of hydroxide as the first step. The conversion of pyrimidine into pyrazole in high yield similarly begins by nucleophilic addition of hydrazine.

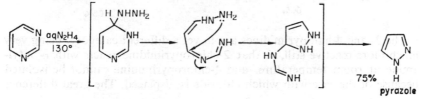

75%   pyrazole

1   SUBSTITUTION WITH HYDRIDE TRANSFER

The diazines readily add Grignard and alkyl and aryllithium compounds

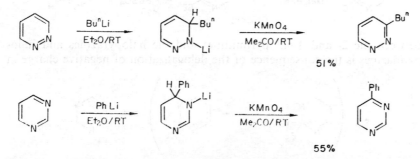

51%

55%

to give non-aromatic products. These frequently require rather more vigorous oxidizing conditions to achieve aromatization by hydride abstraction than do corresponding dihydropyridines: this may be related to the lower resonance energies of the diazines. It is of interest to note that whereas pyridazine reacts with lithium alkyls and aryls at C 3, Grignards react at C 4; pyrimidines are subject to addition to C 4 and not at all to C 2: these orientations are not yet clearly understood.

Sodamide in liquid ammonia easily adds to all three ring systems to give the salts shown below: conversion into the corresponding aromatic amino

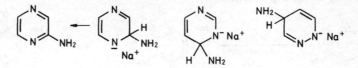

derivative by loss of hydride has only been observed in the case of pyrazine, though 4-methylpyrimidine has been reported to react with $NaNH_2$ at 130–160° C to give the 2-amino derivative.

2   SUBSTITUTION WITH DISPLACEMENT OF HALOGEN

Once again the effect of the extra ring nitrogen is apparent in the nucleophilic displacement of halide by the usual reagents: 2-chloropyridine is appreciably less reactive than 2-chloropyrazine and 3-chloropyridazine.

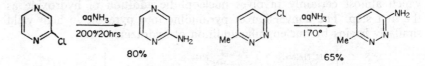

The 2- and 4-halopyrimidines are in yet a different category, being very much more reactive still, so that 2-chloropyrimidine reacts with n-butylamine at room temperature, and 4-chloropyrimidine cannot be isolated because of the ease with which chloride is displaced. The great difference

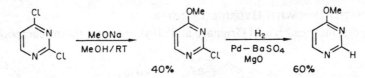

between the 2- and 4-halopyrimidines and the halopyrazines and halopyridazines is the consequence of the delocalization of negative charge in

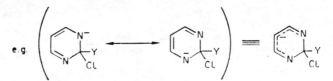

the reaction intermediates over both nitrogens in pyrimidines, a stabilization not accessible to either of the other two diazines. The greater reactivity of 4-chloropyrimidine parallels that of 4-chloropyridine when compared with the 2-isomer.

5-Halopyrimidines are unique among the halodiazines in that the halogen is neither α- nor β- to a nitrogen, these compounds are therefore the least susceptible to nucleophilic displacement.

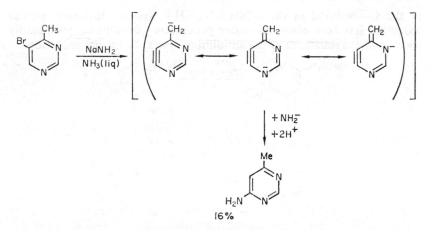

16%

5,6-Pyrimidynes have been postulated as intermediates in the reaction of 4-substituted-5-bromopyrimidines with sodamide: the reaction is complicated by deprotonation of the CH_3, and NH_2^- addition occurs exclusively at C 6.

3   METALLATION OF HALODIAZINES

Halogen-metal interchange has been achieved at very low temperature with 5-bromopyrimidine: at higher temperatures ($- 80°$) addition reactions interfere.

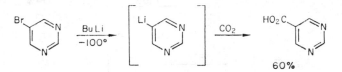

60%

D   Reactions with Reducing Agents

The three diazines are susceptible to reduction under a variety of conditions, and each shows aspects peculiar to itself.

Pyridazine and pyrazine have been reduced only to hexahydroderivatives: both systems are reduced by sodium in boiling ethanol, the pyridazines being partly reduced further to diaminobutanes by N—N fission. Tetramethylpyrazine as free base is not reduced by H_2/Pt, but reduction of the hydrochloride proceeds smoothly to completion.

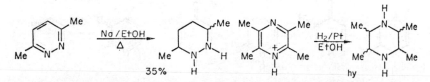

By contrast, pyrimidine does not seem to have been reduced to a hexahydroderivative, probably because of easy hydrogenolysis or hydrolysis of the C—N bond in the —NH.CH$_2$.NH— system. However, partial reduction has been observed under protonating conditions, presumably because of the resistance of the amidinium system to further reduction.

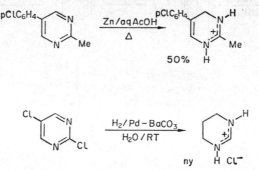

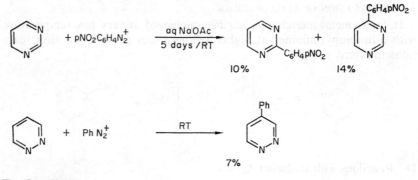

E   Reactions with Free Radicals

The little that is known suggests that C 5 in pyrimidine is perhaps the least susceptible to aryl radical attack, and that in pyridazine C 4 is perhaps the most reactive.

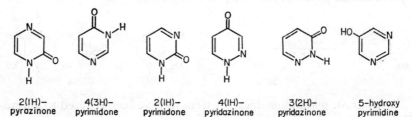

F   Oxydiazines

1   STRUCTURE AND OCCURRENCE

With the exception of 5-hydroxypyrimidine, all mono-oxygenated diazines occur predominantly as the carbonyl tautomers, and thus are categorized as diazinones: 5-hydroxypyrimidine is analogous with β-hydroxypyridine. The structures shown are the non-polarized canonical

| 2(1H)– | 4(3H)– | 2(1H)– | 4(1H)– | 3(2H)– | 5-hydroxy |
| pyrazinone | pyrimidone | pyrimidone | pyridazinone | pyridazinone | pyrimidine |

forms: the molecules actually are resonance hybrids involving the zwitterionic canonical forms (cf. p. 61). Because of the 1,3-relationship of the nitrogens, the two pyrimidones can each exist in two tautomeric forms: 4(3H)-pyrimidone⇌4(1H)-pyrimidone, and 2(1H)-pyrimidone⇌ 2(3H)-pyrimidone.

The dioxydiazines present a more complicated picture, for in some cases where both oxygens are α or γ to nitrogen, and one might expect both to exist in carbonyl forms, one exists as hydroxyl. A well-known example is maleic hydrazide, a widely used selective plant-growth regulator (3-hydroxy-6(1H)-pyridazinone), which could well exist in the dicarbonyl form but which prefers to exist as the hydroxypyridazinone: very probably a reason for this is that the diamide tautomer is destabilized by the juxtaposition of the two partially-positive nitrogens.

A different and quite special case is that of 4,6-dioxypyrimidine which could exist as 4,6(1H, 5H)-pyrimidione or as 4-hydroxy-6(1H)-pyrimidone but actually as the meso-ionic system, 1, which may be looked upon as a combination of amidinium and β-ketoenolate anion (see p. 324).

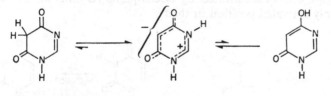

4,6(1H,5H)−pyrimidione   1   4−hydroxy−6(1H)−pyrimidone

On the other hand, uracil exists as 2,4(1H, 3H)-pyrimidione and is usually written as 2, but one has to remember that such amide groups are mesomeric and highly polarized: a more correct formulation is 3. Uracil is

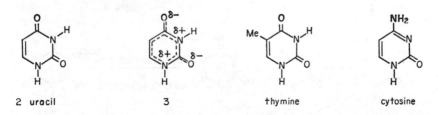

2 uracil   3   thymine   cytosine

biologically one of three extremely important pyrimidines present in nucleotides (see p. 326): the other two are thymine and cytosine.

The 5,5-disubstituted derivatives of barbituric acid, barbiturates are well-known chemotherapeutic agents.

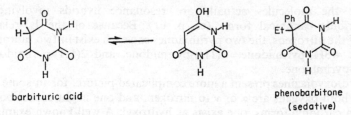

barbituric acid

phenobarbitone
(sedative)

2   REACTIONS WITH ELECTROPHILIC REAGENTS

One oxygen substituent is not always sufficient to allow electrophilic substitution to occur: thus 3-pyridazinone can be neither nitrated nor halogenated, and of the monopyrimidones only 2(1H)-pyrimidone can be nitrated; pyrazinones appear to be the least resistant to electrophilic substitution.

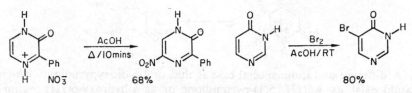

The dioxypyrimidines uracil and 4,6-dioxypyrimidine are much more reactive: the latter can even be nitrosated. 2-, 4-, or 6-mono- and dioxypyrimidines are always substituted by electrophilic reagents at C 5 since this is the only activated position in them.

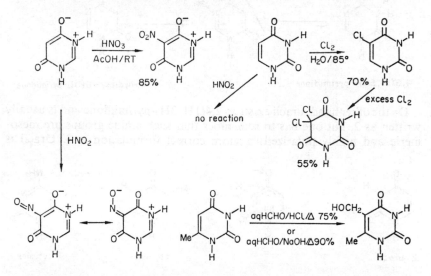

5-Hydroxypyrimidine, the only phenolic diazine, is very surprisingly decomposed even by dilute acids; no electrophilic substitutions have been reported.

Diazinones react easily with phosphorus oxychloride and related reagents to give corresponding chlorodiazines, exactly as pyridones yield chloro-pyridines: this reaction involves initial electrophilic attack at carbonyl oxygen (see p. 63).

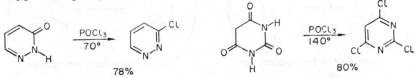

3 DEPROTONATION

Diazinones are deprotonated in much the same way as pyridones to give mesomeric anions. These anions are ambident, and may be alkylated

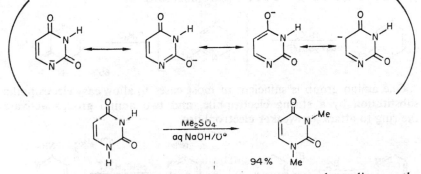

or acylated predominantly on nitrogen, or on oxygen, depending on the reaction conditions and the particular diazinone. The dioxypyrimidines react by way of their anions with aryldiazonium ions and with formaldehyde at C 5.

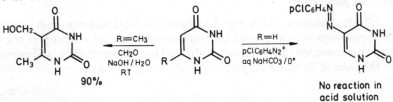

4 CARBENE ADDITION

Carbene itself and the dihalocarbenes add in non-hydroxylic solvents, but not in water, to the C5-C6 double bond of several N-dialkylated uracil and thymine derivatives.

G Aminodiazines

Aminodiazines exist in the amino form. They are stronger bases than the corresponding unsubstituted systems and always protonate on one of the ring nitrogens: where two isomeric cations are possible, the order of preference for protonation is of a ring nitrogen which is $\gamma > \alpha > \beta$ to the NH_2 group, as is seen in the two examples given. A corollary of this is that those amino diazines which contain a γ-aminopyridine system are the strongest bases.

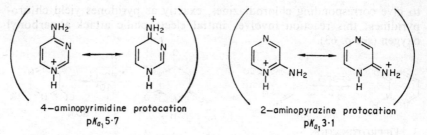

4-aminopyrimidine protocation   2-aminopyrazine protocation
        pK_{a_1} 5·7   pK_{a_1} 3·1

All aminodiazines react with nitrous acid to give the corresponding diazinone by way of the highly reactive diazonium salt. Even 5-amino pyrimidine does not give a stable diazonium salt.

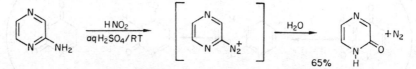

One amino group is sufficient in most cases to allow easy electrophilic substitution by a strong electrophile, and two amino groups activate the ring to attack by weaker electrophiles.

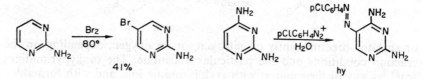

H   Alkyldiazines

All alkyldiazines, with the exception of 5-alkylpyrimidines, undergo condensations which involve deprotonation of the alkyl group as the first step: the intermediate anions are stabilized by mesomerism involving one or, in the case of 2- and 4-alkylpyrimidines, both nitrogens.

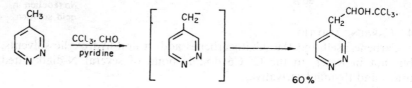

In pyrimidines, a 4-alkyl group is more readily deprotonated than a 2-alkyl: here again we see the greater stability of a γ-quinoid mesomeric system.

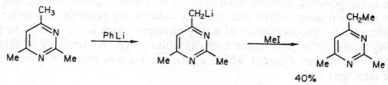

I Diazine N-Oxides

The main body of work here has been concerned with the pyrazine and the pyridazine N-oxides: relatively little has appeared on the use of pyrimidine N-oxides. Indeed, the preparation of the N-oxide of pyrimidine leads to extensive oxidative breakdown; alkyl and alkoxypyrimidines can, however, be N-oxidized normally.

Pyridazine N-oxide can be produced normally, and undergoes nitration in much the same way as does pyridine N-oxide.

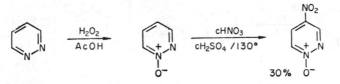

Nucleophilic substitution of halogen occurs much more readily in N-oxides than in the parent halodiazines, even when substitution is

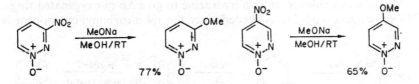

occurring β to the N-oxide grouping. This β-activation is also seen in the displacement of nitrite by methoxide in 3-nitropyridazine-1-oxide where it occurs as readily as in the 4-nitro isomer.

Nucleophilic substitution with concomitant loss of the oxide function also occurs as smoothly with the diazine oxides as with the pyridine oxides.

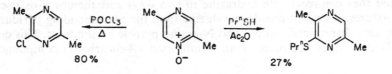

J Synthesis of Diazines

1 SYNTHESIS OF PYRIDAZINE RINGS

The only generally important method for the preparation of the pyridazine ring involves the reaction of a 1,4-dicarbonyl compound with hydrazine (or a substituted hydrazine). The most useful procedure makes use of

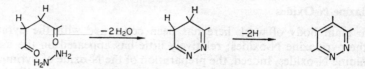

a 4-keto-ester. This gives rise to a dihydropyridazinone which can be easily

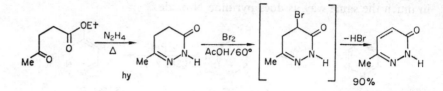

oxidized to the fully-aromatic heterocycle, (often by bromination and de-hydrobromination by virtue of the activation provided by the carbonyl group).

Maleic anhydrides react with hydrazine to give 3,6-di-oxygenated rings. No subsequent oxidation is needed since the 1,4-dicarbonyl component is

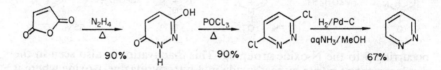

unsaturated to begin with. The best laboratory synthesis of pyridazine itself makes use of this route.

Saturated 1,4-diketones, in this approach, suffer from the disadvantage that they can react with hydrazine in two ways and thereby give rise to mixtures of desired dihydropyridazine (and the corresponding pyridazine by easy disproportionation) and N-amino pyrrole by-products. This complication does not occur if an *unsaturated* 1,4-dicarbonyl compound is

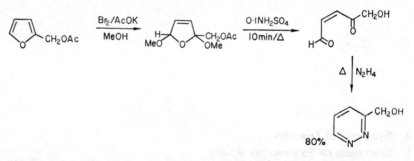

used. Such compounds are readily available by oxidation of furans (see p.242) and react with hydrazines directly to give fully aromatic pyridazines.

2 SYNTHESIS OF PYRIMIDINE RINGS

The most useful and the most-used method for the construction of a pyrimidine ring from non-heterocyclic precursors involves the fusion of two three-atom units. The two components can be generally designated as

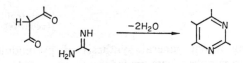

1,3-dicarbonyl component and a *di-amino* component. Pyrimidine can be obtained in the laboratory from uracil in two steps, the synthesis of which illustrates this general approach. The 1,3-dicarbonyl component in this

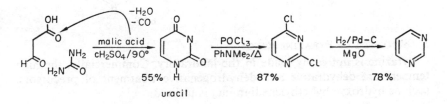

example is α-formylacetic acid (itself produced *in situ* by decarbonylation of malic acid with concentrated sulphuric acid) and the di-amino component is urea.

Usually it is not known in what precise order the two nucleophilic additions and eliminations of water (or alcohol) occur. In an exceptional example an intermediate was isolated in which only one of the two condensations had occurred.

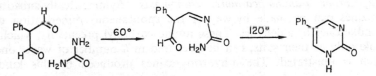

Thiourea, imino-ethers, amidines and guanidines can be used instead of urea in this approach. Indeed, reactivity in this synthetic procedure is related to the nucleophilicity of the amino groups, and the list above represents an increasing order of reactivity.

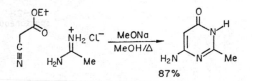

Barbiturate drugs carrying two substituents at C 5 can also be made easily by this kind of synthesis.

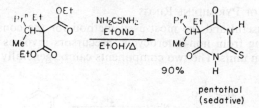

90%

pentothal
(sedative)

The following recent general synthesis of pyrimidines is a simple variant, and in its simplest form provides a good route to pyrimidine itself.

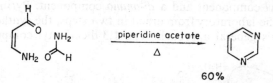

60%

3 SYNTHESIS OF PYRAZINE RINGS

Pyrazine is not easily made in the laboratory. Commercially, the high temperature dehydrative and dehydrogenative treatment of precursors, such as hydroxyethyl ethylenediamine, is utilized.

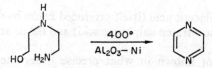

Two general methods are available for the preparation of pyrazine rings.

(a) *From α-amino carbonyl compounds.* Symmetrically-substituted pyrazines can be made by way of the spontaneous dimerization of α-aminoketones or aldehydes. These amino-carbonyl precursors, which are stable only as their salts, can be prepared in a number of ways, one of which is illustrated. The dihydropyrazines produced by this kind of

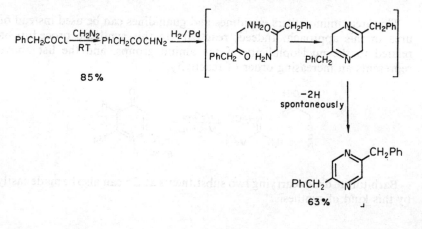

approach are very easily dehydrogenated, and often distillation alone is sufficient to bring about disproportionation.

α-Amino carboxylic esters dimerize easily to 2,5-diketopiperazines, which are stable and not prone to disproportionation.

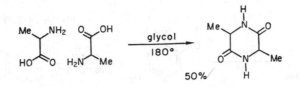

(b) *From 1,2-dicarbonyl compounds and 1,2-diamino compounds.* In this general method the pyrazine ring is constructed by condensing a 1,2-dicarbonyl compound with an appropriate diamine, with subsequent dehydrogenation if necessary.

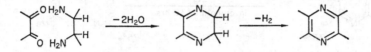

Many derivatives of 2-aminopyrazine have been made for biological evaluation, for 2-sulphonamidopyrazine (sulphapyrazine) is a powerful bacteriostat. An ingenious modification of the general method for constructing pyrazine rings has been used to make many of these compounds. This involves using 5,6-diamino-2,4-(1H, 3H)-pyrimidione as the diamino component: the products can be hydrolysed with cleavage of the pyrimidine ring to give 2-aminopyrazines.

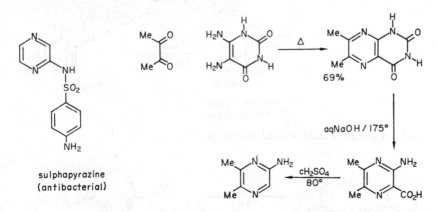

sulphapyrazine
(antibacterial)

4   EXAMPLES OF DIAZINE SYNTHESES

(a)   *4,6-Diamino-5-thioformamido-2-methyl pyrimidine*, and conversion into 2-methyl adenine.

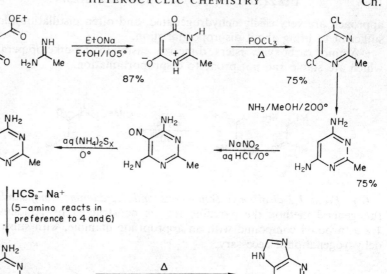

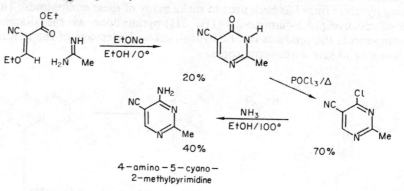

35%　　　　　　　　　　　　　　　　　　　　　　　　75%

4,6-diamino -5-thioformamido-
2-methylpyrimidine

2-methyladenine

(*b*)　*4-Amino-5-cyano-2-methyl pyrimidine*, intermediate in the synthesis of vitamin B$_1$ (see p. 320).

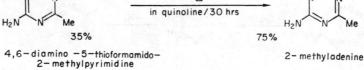

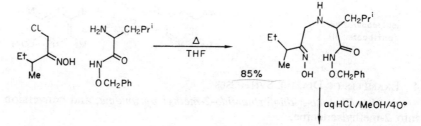

20%

40%　　　　　　　　　　　　　70%

4-amino-5-cyano-
2-methylpyrimidine

(*c*)　*Aspergillic acid*, mould metabolite.

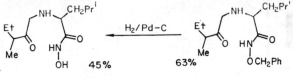

aspergillic acid

Pyrylium Salts and Pyrones: General Discussion and a Comparison with Pyridinium Salts, Thiopyrylium Salts and Pyridones

Pyrylium Salts

pyrylium cation

The pyrylium cation falls into the category of aromatic systems which are so reactive that they appear to be unstable; other well-known examples in this category are tropylium cation and cyclopentadienyl anion (see p.4). It may at first sight seem strange that molecules which are said to be stabilized by aromatic resonance react extremely readily to give less reactive non-aromatic products, as is seen in the reaction of pyrylium with nucleophiles to give pyran derivatives.

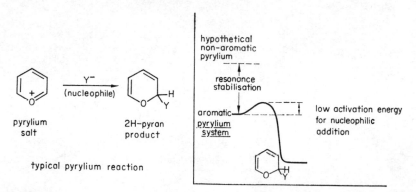

pyrylium salt → 2H-pyran product

typical pyrylium reaction

hypothetical non-aromatic pyrylium

resonance stabilisation

aromatic pyrylium system

low activation energy for nucleophilic addition

However, one must appreciate that stability and reactivity are relative attributes: thus, the degree of reactivity of pyrylium cation is such as to allow its isolation at ordinary temperatures in appropriately weakly nucleophilic solvents, and this isolability is the consequence of aromatic resonance stabilization: put into other words, a hypothetical pyrylium cation not stabilized by aromatic resonance would very probably be so reactive as to make its isolation impossible.

Here, it might be appropriate to point out the difference between reactivity and stability: pyrylium cation is a relatively *stable* system, for the crystalline perchlorate does not decompose below 275°, but it is a very *reactive* system towards nucleophiles, for it reacts rapidly with water at room temperature. A truly unstable compound might be one which, say, thermally decomposes at ordinary temperatures even as a pure substance, as for example the molozonide of tetramethylethylene.

Almost all the known reactions of the pyrylium ring-system involve extremely easy addition of a nucleophile to an α-position as the first step, often followed by further processes (see p. 149ff. for details): the closely analogous N-alkylpyridinium cations similarly add nucleophiles, but except for reactions such as alkyllithium additions, in which reversal of addition

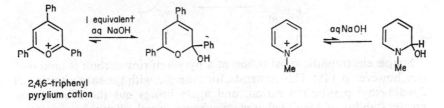

2,4,6-triphenyl
pyrylium cation

is not possible, effectively the equilibrium is entirely on the side of the aromatic pyridinium cation. Thus, where triphenylpyrylium is completely converted into the hydroxyl adduct by mild base, N-methylpyridinium cation which carries no stabilizing substituents is unaffected. Pyridinium salts carrying powerful electron-withdrawing groups on nitrogen behave more like the pyrylium systems, and undergo ring opening (see p. 70); the substituent on the nitrogen reduces the stability of the salt with respect to the product of nucleophilic addition.

This marked difference in the positions of comparable equilibria is the consequence of the greater electronegativity of oxygen compared with

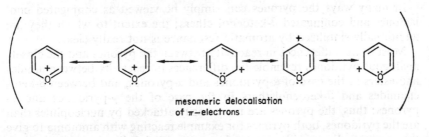

mesomeric delocalisation
of π-electrons

inductive polarisation
of σ skeleton, which
mainly affects α-carbons

nitrogen: thus oxygen tolerates a positive charge much less readily, and hence in pyrylium the α- and γ-ring carbons carry a much greater fractional positive charge than pyridinium α- and γ-carbons and are much more electrophilic.

Although most of the reactions studied have involved 2,4,6-trisubstituted pyrylium salts, the generalization can be made that nucleophilic addition occurs preferentially at C 2: this contrasts with the behaviour of halopyridines which react almost equally readily at the α- and γ-positions; and closely resembles the behaviour of quaternary pyridinium salts, in which the additional inductive effect of the positive charge, more powerfully felt at the α-positions. strongly directs nucleophiles to those positions.

As in pyridine chemistry, addition of a nucleophile to a β-position does not occur and for the same reasons (see p. 39). It is also useful and instructive to note the analogy between the reactions of pyrylium and those of protonated carbonyl compounds:

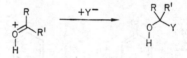

Simple electrophilic substitution at a pyrylium ring carbon is unknown (see, however, p. 148). This contrasts, for example, with the easy nitration of 2,6-dimethyl pyridinium cation, and again brings out the effect of the greater stability of positive charge on nitrogen, which allows the generation of a doubly-charged reaction intermediate.

α- And γ-Pyrones

In many ways the pyrones can simply be viewed as conjugated enol lactones and conjugated β-ketoenol ethers; the extent to which they are additionally stabilized by aromatic resonance is not really clear.

Most of the differences in reactivity between the pyrones and the closely analogous pyridones resemble the differences in reactivity between amides and esters in the case of α-pyridones and α-pyrones, and between β-ketoenamides and β-ketoenolethers in the case of the γ-pyridones and γ-pyrones: thus, the pyrones are more easily attacked by nucleophiles than are the pyridones, both pyrones for example reacting with ammonia to give the corresponding pyridones, much as esters and β-dicarbonyl derivatives react with ammonia to give amides and β-ketoenamides. That the pyridones cannot be converted into the pyrones by alkali illustrates the greater stability of the former. Furthermore, the greater basicity of γ-pyridone compared with γ-pyrone is reflected in the basicity differences between the appropriate acyclic analogues.

Finally, it is interesting to note that the lower delocalization energy of the pyrones shows up in the formation of dibromo-adducts, not observed in the pyridones.

Thiopyrylium Salts

The thiopyrylium cation is stabler than pyrylium; thus thiopyrylium iodide can be recrystallized from hot water.

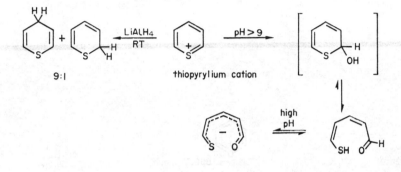

Reaction with hydroxide ion, however, does occur as shown. The sulphur system is susceptible to powerful nucleophiles just as is pyrylium, and lithium aluminium hydride, for example, attacks at C 2 and C 4.

The thiopyrylium system must be distinguished from the thiabenzene nucleus, in which sulphur is formally tetravalent and the molecule as a whole is neutral. 1-Methyl-3,5-diphenylthiabenzene has been shown to be a non-aromatic ylid.

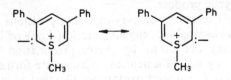

12

Pyrylium Salts and Pyrones: Reactions and Synthesis

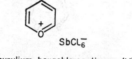

pyrylium hexachloroantimony (V)

Pyrylium salts, especially the perchlorate and the hexachloroantimony (V) salts, are stable but reactive compounds which so far have not been found in living organisms, even though the benzopyrylium system plays such an important role in the flower pigments. Nor have pyrylium salts found their way into chemotherapy.

Almost all the known reactions of pyrylium salts involve addition of a nucleophile to C 2, or sometimes to C 4, as the first step: it depends on the nature of the substituents on the ring and on the particular nucleophile involved whether the initial addition product may be isolated as such or go on to react further. A feature of pyrylium salt chemistry is ring-opening of the adduct, followed by cyclization in a different way to give a new hetero-cyclic or homocyclic product.

Straightforward reactions of pyrylium ring positions with free-radical reagents, or with electrophilic reagents, are not known: H-D exchange at C 3 and C 5 may not occur by direct protonation of the pyrylium cation but rather by way of a neutral intermediate formed by addition of acetate (see p. 149). Controlled oxidations, like that of pyridinium cations to 2-pyridones, are likewise not known in pyrylium chemistry.

A Reactions with Nucleophilic Reagents

(a) *Water and hydroxide ion.* The degree of susceptibility of pyrylium salts to nucleophilic attack varies widely, from the extremely reactive unsubstituted pyrylium cation itself which is even attacked by water at

$0°$, to 2,4,6-trimethylpyrylium cation which is stable in water at $100°$: here again we see what a remarkably powerful stabilizing effect methyl groups exert in cations.

Hydroxide anion, however, adds very readily to C 2 in all cases: with 2,4,6-triphenylpyrylium cation the resulting 2-hydroxy-2H-pyran, which is a cyclic enol hemiacetal, is in equilibrium with a dominant concentration

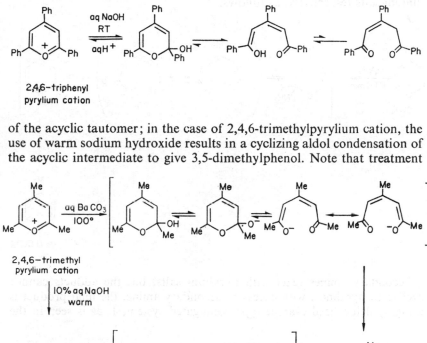

2,4,6–triphenyl
pyrylium cation

of the acyclic tautomer; in the case of 2,4,6-trimethylpyrylium cation, the use of warm sodium hydroxide results in a cyclizing aldol condensation of the acyclic intermediate to give 3,5-dimethylphenol. Note that treatment

2,4,6–trimethyl
pyrylium cation

of the acyclic unsaturated diketones with acid regenerates the original pyrylium salt.

(b) *Acetic acid.* 2,4,6-Triphenyl pyrylium reacts with hot O-deutero-acetic acid (CH_3CO_2D) to give the 3,5-dideuterated product: this probably

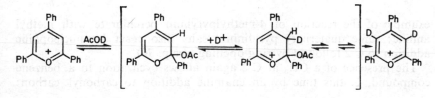

occurs by initial nucleophilic addition of acetate to C 2 as shown. 2,4,6-Trimethyl and 2,6-dimethylpyrylium are likewise deuterated at C 3 and C 5, but even more rapidly at the methyls (see p. 153).

(c) Ammonia, primary, and secondary amines. Ammonia and primary amines react with pyrylium salts to give pyridines and quaternary pyridinium salts respectively, as follows.

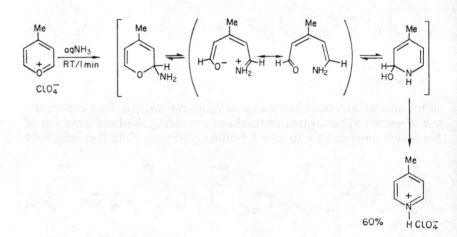

Secondary amines react with pyrylium salts, but the adducts cannot cyclize to pyridines: with excess of secondary amine, the final product is a highly delocalized cyanine-type conjugated system, 1, as is seen in the

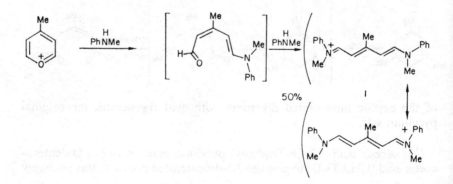

example of the reaction of 4-methylpyrylium perchlorate with methyl aniline: some quaternary pyridinium salts also react by nucleophilic addition to C 2 followed by ring opening (see p. 70).

The presence of a CH_3 at C 2 again allows cyclization to a benzene compound, 2, this time by an enamine addition to carbonyl carbon:

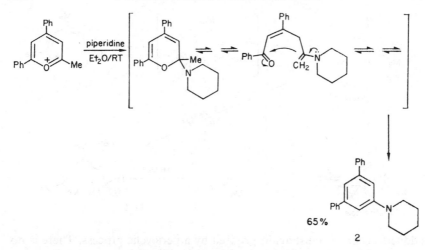

65%

2

(d) *Nitromethane anion.* Once again, a series of equilibria initiated by nucleophilic addition of nitromethane anion to C 2 leads, eventually, by way of an intramolecular aldol condensation, to 3.

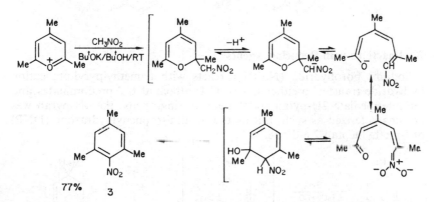

77%   3

(e) *Cyanide ion.* Addition to C 2 in trimethylpyrylium leads to non-isolable intermediate, 4, which almost certainly ring-opens by a pericyclic process (that is, by way of a cyclic non-polar transition state).

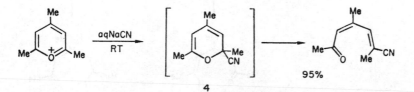

95%

4

(f) *Organometallic reagents.* These add to C 2, and sometimes to C 4 if that position is unsubstituted. When the ring-opening of the 2H-pyran

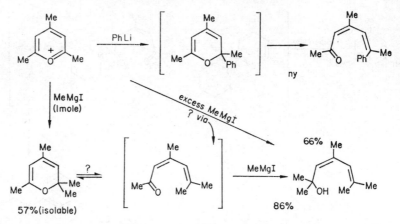

adducts occurs, it must again proceed by a pericyclic process. There is no ring-opening of 4H-pyrans derived by C 4 addition.

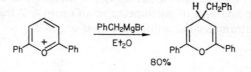

B Reactions with Reducing Agents

Sodium borohydride ($NaBH_4$) reacts with trimethylpyrylium cation by hydride transfer to either C 2 or C 4: attack at C 2 predominates, and the intermediate 2H-pyran spontaneously ring-opens; the 4H-pyran was not characterized as such, but as the bis-dinitrophenylhydrazone (DNP) of 4-methylheptan-2,6-dione.

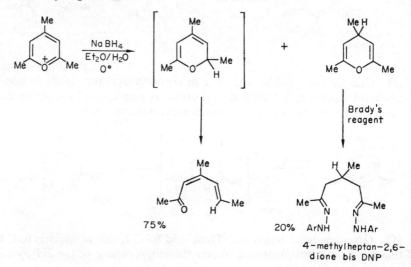

Zinc metal in water reduces pyrylium to give a dimeric 4H-pyran derivative, 5, quantitatively: this presumably proceeds by one-electron transfer followed by radical dimerization (see reduction of pyridine, pp. 59, 60).

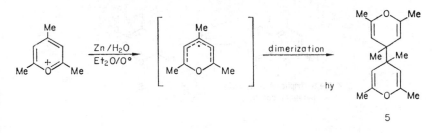

5

C   Alkylpyrylium Salts

Alkyl groups at the α- and γ- positions of pyrylium salts are, as might be expected, quite acidic: thus the C4 methyl of trimethylpyrylium is completely deuterated in D_2O at 100° in ten minutes, and the C 2 and C 6

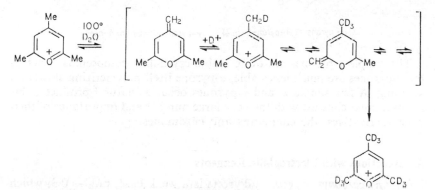

methyls in five hours. The reason for this rate difference is probably associated with the greater stability of the paraquinoid-type intermediate.

H-D exchange in the less nucleophilic AcOD is much slower: under these conditions the ring hydrogens at C 3 and C 5 in 2,6-dimethylpyrylium have been shown to exchange at one-seventeenth the rate of the methyl hydrogens, and the C 4 hydrogen not to exchange at all.

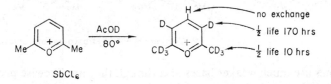

D   Oxypyrylium Salts

2- and 4-hydroxypyrylium salts are quite strongly acidic, and are much better known as their conjugate bases, the α- and γ-pyrones (see below).

Very little is known about 3-hydroxypyrylium salts: one of the very few which have been prepared and characterized, 2,4,6-triphenyl-3-hydroxy-pyrylium chloride, has been isolated as the deep red zwitterion, 6.

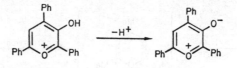

2,4,6−triphenyl−3−hydroxy
pyrylium cation

6

REACTIONS OF α-PYRONES AND γ-PYRONES

α− pyrone (2 H−pyran−2−one)   γ−pyrone (4H−pyran−4−one)

The simple γ-pyrones are quite stable crystalline compounds, whereas the α-pyrones are much less stable, α-pyrone itself polymerizing slowly on standing. A few simple α- and γ-pyrones occur as natural products; this forms a great contrast with the very large number and importance of their benzo-derivatives, the coumarins and chromones.

A   Reactions with Electrophilic Reagents

(a)   At carbonyl oxygen.   γ-Pyrone is a weak base, $pK_a - 0.3$, which is protonated on the carbonyl oxygen: crystalline salts may be obtained, generally by working in a non-hydroxylic solvent.

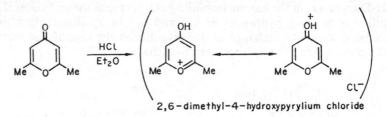

2,6−dimethyl−4−hydroxypyrylium chloride

α-Pyrones are much weaker bases and though they are likewise protonated on carbonyl oxygen in strong acids, salts cannot be isolated.

Acid-catalysed H-D exchange in γ-pyrone occurs at C 3 and C 5, and is believed to proceed by way of the acyclic β-tricarbonyl intermediate, 7, though the alternative pathway shown would seem to be equally plausible; no exchange occurs at C 2 and C 6.

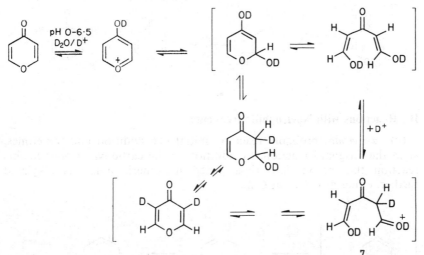

γ-Pyrone, but not α-pyrone, may be alkylated at the carbonyl oxygen by dimethyl sulphate: this reflects the greater degree of polarization of the

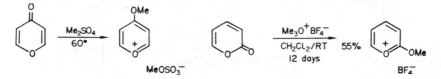

γ-system. However, α-pyrone does react with the more powerful trimethyl-oxonium borofluoride (Meerwein reagent).

(b) *At ring carbon.* Both α- and γ-pyrones undergo electrophilic substitution at C 3 and C 5, positions *o-* or *p-* to the carbonyl group. In its reaction with $NO_2{}^+BF_4{}^-$, α-pyrone rapidly and reversibly forms an O-nitro salt, and is converted more slowly into 5-nitro-α-pyrone.

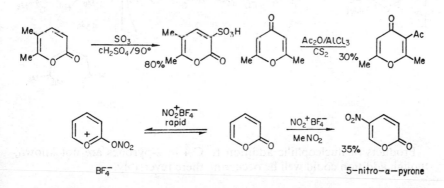

In some instances reaction with halogens proceeds by an addition-elimination process:

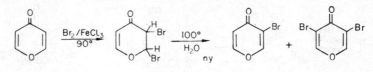

B Reactions with Nucleophilic Reagents

(*a*) *α-Pyrones* probably react by initial C 6 addition and sometimes, as in the Grignard reaction, by addition to the carbonyl carbon: in this reaction the intermediate ring-opened mesomeric anion is alkylated further, either at C 6 or at C 2.

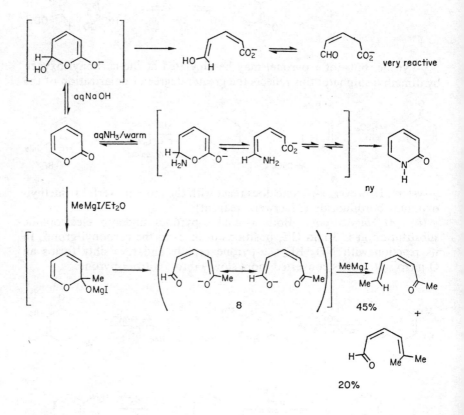

Products of nucleophilic addition to C 4 in α-pyrones are not known, though addition could well be occurring there reversibly.

(*b*) *γ-Pyrones* react similarly, with nucleophilic reagents, they are thus converted in good yields into γ-pyridones by ammonia or primary amines.

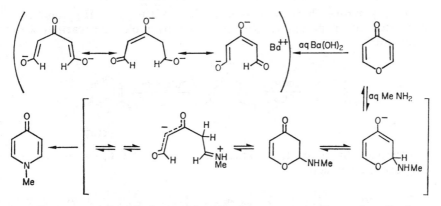

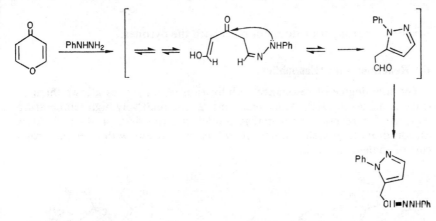

γ-Pyrones react with the usual carbonyl reagents such as phenylhydra-zine, not, however, in the simple manner at the carbonyl group but with ring-opening and further reaction:

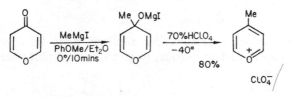

Grignard reagents, as with α-pyrones, react with the carbonyl carbon, but in this case the product cannot ring-open and can be converted in high yield into 4-monosubstituted pyrylium salts, which thus provides a good and easy synthesis of these otherwise relatively inaccessible pyrylium salts.

C   Reactions with Reducing Agents

Catalytic reduction of α- and γ-pyrones occurs first at the C—C double bonds: α-pyrones give saturated δ-lactones or pentanoic acids by C 6-O hydrogenolysis.

Some γ-pyrones have been reduced by LAH to give 4H-pyran-4-ols, converted into pyrylium salts by acid: this resembles the reaction with Grignards.

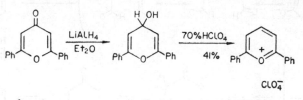

α-Pyrones, by contrast, are attacked by LAH at C 6, which leads to ring-opened products:

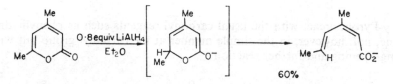

Sodium borohydride does not react with the pyrones.

D   Reactions with Dienophiles

The low degree of resonance stabilization of α-pyrones allows them to react as dienes in Diels-Alder reactions: at the relatively high temperature required for reaction with maleic anhydride, the first product, 9, loses carbon dioxide to yield a diene, 10, which then reacts with a second molecule of maleic anhydride.

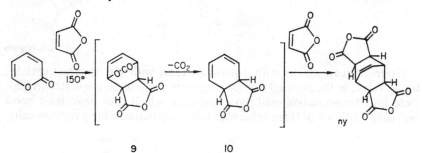

Dimethyl acetylene dicarboxylate similarly adds to α-pyrones, with subsequent loss of carbon dioxide, to give dimethyl phthalate derivatives

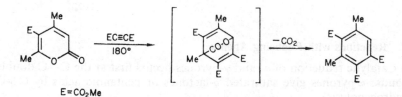

E = CO_2Me

which have been of value in determining the structure of naturally-occurring α-pyrones.

There seems to be no record of a γ-pyrone serving as a dienophilic component in a Diels-Alder reaction.

THE SYNTHESIS OF PYRYLIUM SALTS AND OF THE α- AND γ-PYRONES

The synthesis of these systems involves the formation and cyclization of an appropriately unsaturated 1,5-dicarbonyl precursor: the general scheme shows how the formation of any one of these three types depends on the oxidation pattern of this precursor:

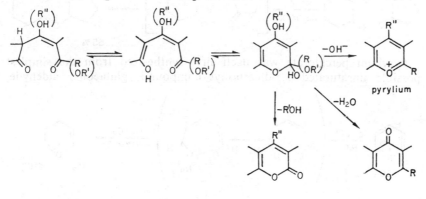

α−pyrone   γ−pyrone

The cyclization step is simply hemiacetal formation by the enolic OH, and loss of water or alcohol completes the synthesis.

In the great majority of syntheses, the 1,5-dicarbonyl precursor is generated in the reaction mixture by an aldol, Claisen, or Michael condensation between two appropriate reactants.

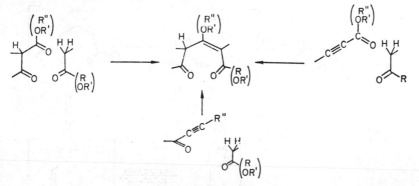

Another approach has been the 1,3-diacylation of a unit representing C 3, C 4, and C 5 of the final ring.

A Pyrylium Salt Synthesis

An example of the simplest type of synthesis from a ketone and a 1,3-dicarbonyl compound is shown below:

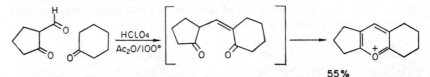

55%

Pyrylium perchlorate was itself first synthesized from the simplest possible unsaturated 1,5-dicarbonyl compound, glutaconic aldehyde,

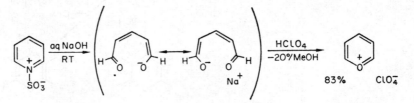

83% ClO_4^-

very conveniently prepared from the pyridine sulphur trioxide compound as shown.

A general pyrylium synthesis which does not fall within the above generalizations, but which is closely analogous with the Hantzsch pyridine synthesis, must be mentioned: this leads to a pyran, which has to be oxidized to the pyrylium system.

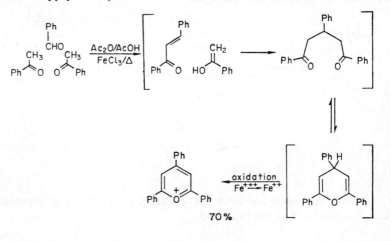

70%

The following synthesis of 2,6-diphenylpyrylium chloride has been claimed to involve a [4+ + 2] cycloaddition.

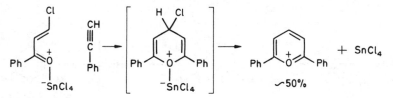

B α-Pyrone Synthesis

A simple example of the general method is the synthesis of coumalic acid by the self-condensation of formyl acetic acid, itself generated by the action of sulphuric acid on malic acid.

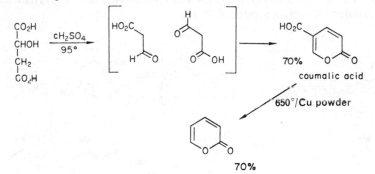

coumalic acid

650°/Cu powder

70%

C γ-Pyrone Synthesis

A general route is illustrated by the Claisen condensation of ethyl phenylpropiolate with acetone:

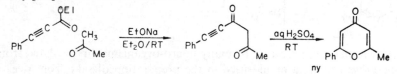

ny

Another example illustrates the 1,3-diacylation type of approach:

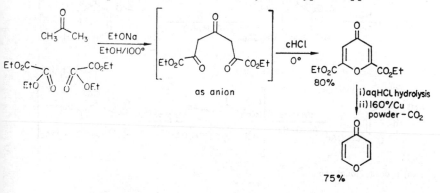

as anion

80%

i) aqHCl hydrolysis
ii) 160°/Cu powder − CO$_2$

75%

The simplest γ-pyrone synthesis is that of 2,6-dimethyl-γ-pyrone from Ac_2O and polyphosphoric acid.

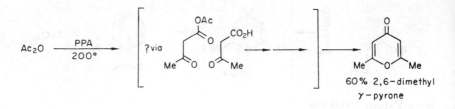

D   Examples of Pyrylium and Pyrone Synthesis

(a) *4-Ethyl-2,6-dimethyl pyrylium and 2,4,5,6-tetramethyl pyrylium salts,* by acylation of olefins produced *in situ*.

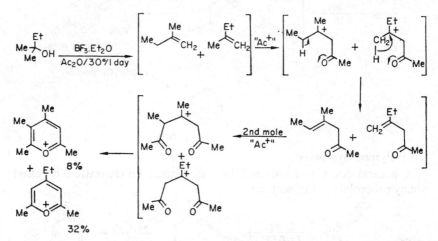

(b) *Kojic acid,* produced by many micro-organisms, first isolated from *Aspergillus oryzae,* a mould used in the preparation of sake from rice.

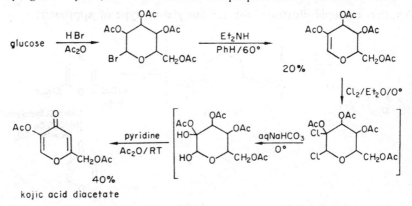

(*c*) *Yangonin,* isolated from Polynesian kawa shrub, *Piper methysticum,* which is used to make a local intoxicating drink.

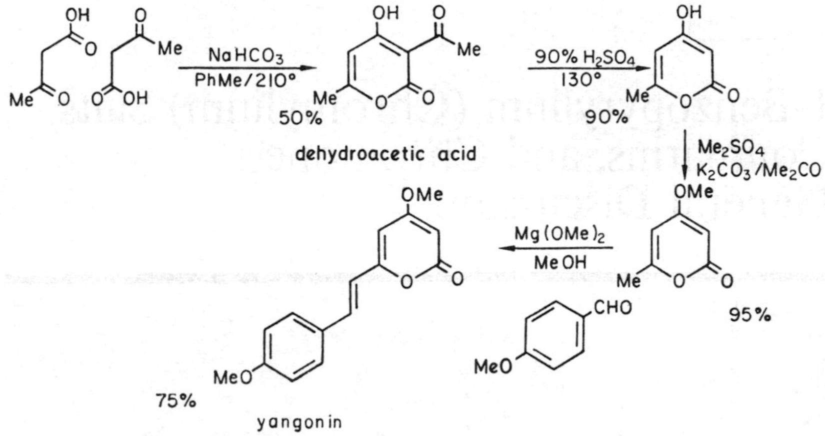

dehydroacetic acid

yangonin

1-Benzopyrylium (Chromylium) Salts, Coumarins, and Chromones: General Discussion

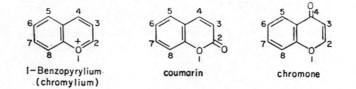

| 1–Benzopyrylium (chromylium) | coumarin | chromone |

Nucleophilic addition to C 2 is the main, almost the only, type of reaction known to be undergone by 1-benzopyrylium, for there seems to be no reported electrophilic substitution reactions involving the homocyclic ring. This failure to react with electrophiles again contrasts sharply with the many electrophilic substitution reactions undergone by quinolinium cations under relatively mild conditions (see p. 89), and further underlines the much more powerful deactivating effect of positive oxygen when compared with positive nitrogen.

The coumarins and chromones react with both nucleophiles and electrophiles, much in the same way as do the two quinolones.

Note that the number in 1-benzopyrylium refers to the position of the oxygen: the isomeric 2-benzopyrylium system (p. 168) is the analogue of isoquinoline.

1-Benzopyrylium (Chromylium) Salts, Coumarins, and Chromones: Reactions and Synthesis

These three ring systems are very widely distributed throughout the plant kingdom, where many hundreds (even thousands) of very varied secondary metabolites contain them. Not the least of these compounds are the anthocyanins and flavones, which make up the majority of the flower pigments; in addition, many flavone and coumarin derivatives in plants have marked toxic and other physiological properties in animals. It is remarkable, however, that these systems play no part in the normal metabolism of animals.

Chemotherapeutically valuable compounds in this class are a series of coumarins, of which acenocoumarol is one, which are very valuable as anti-coagulants, and intal, which is used in the treatment of bronchial asthma.

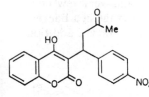 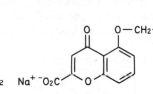 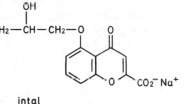

intal

acenocoumarol
(anticoagulant)

REACTIONS OF 1-BENZOPYRYLIUM AND FLAVYLIUM (2-PHENYL-1-BENZOPYRYLIUM) SALTS

By far the most important aspect of the reactivity of the 1-benzopyrylium system is the ease with which nucleophiles add to C 2 and C 4; in many ways, as with pyrylium salts, the reactions are analogous with those of O-protonated carbonyl compounds.

1-benzopyrylium
(chromylium) ion

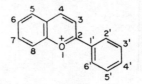

flavylium (2-Phenyl-1-benzopyrylium)ion

No examples are known of electrophilic or radical substitution of either the homocyclic or heterocyclic rings of 1-benzopyrylium salts: flavylium salts can be nitrated, but only in the isolated benzene ring.

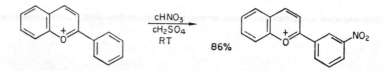

The 1-benzopyrans (chromenes) will not be discussed, except insofar as they are products of reaction.

2H-1-Benzopyran
(2H-chromene)

4H-1-Benzopyran
(4H-chromene)

A Reactions with Oxidizing Agents

Most oxidizing agents lead to general breakdown and very few controlled oxidations are known: one of these is C 2-C 3 cleavage in a Baeyer-Villiger type of reaction, in which the first step is likely to be C 2 addition of peroxide.

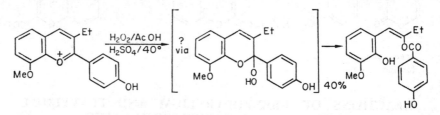

B Reactions with Nucleophilic Reagents

(a) *Water and alcohols.* Water and alcohols add under controlled conditions either as such or as OH⁻ or OR⁻ to give chromenols or chromenol ethers; addition occurs to C 2 or C 4 depending on the substitution pattern.

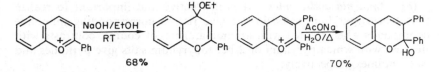

68%   70%

More vigorous alkaline treatment of 1-benzopyrylium salts results in ring-opening and subsequent C—C bond cleavage by a retro-aldol mechanism as shown:

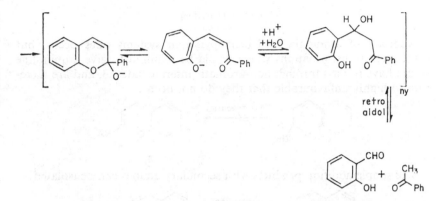

An alternative C—C cleavage is open to 1-benzopyrylium salts carrying a phenolic hydroxyl at C 7; this extra oxygen, in the appropriate intermediate anion, 1, allows a sufficiently high rate of C-protonation of the benzene ring to an intermediate such as 2 to make a retro-aldol type of reaction possible, leading to resorcinol and the four possible fission products of the β-diketone.

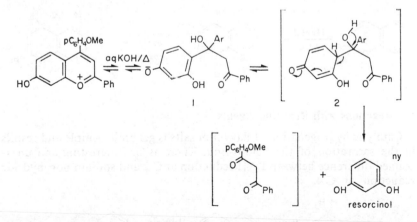

These reactions are essentially the reverse of the syntheses (see p. 178) and were much used in the earlier days of the elucidation of the structure of anthocyanin flower pigments (see p. 169).

(*b*) *Ammonia and amines.* It is instructive and important to realize that 1-benzopyrylium salts do *not* give rise to corresponding quinoline compounds when treated with ammonia. This, of course, contrasts with the ease with which pyrylium or 2-benzopyrylium salts give pyridines and isoquinolines respectively.

2–benzopyrylium
cation

Addition of ammonia to 1-benzopyrylium systems does occur, but subsequent transformations which would be required to give a quinoline would have to pass through non-aromatic intermediates, 3, and are therefore so highly unfavourable that they do not occur.

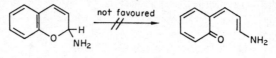

3

The simple addition products with secondary amines can be isolated:

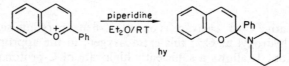

hy

(*c*) *Reactions with carbon nucleophiles.* Grignards add to C 2, and a wide range of milder carbon nucleophiles such as enolates or phenols add to C 4.

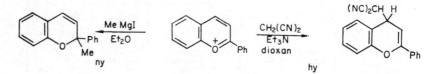

ny   hy

C   Reactions with Reducing Agents

Catalytic hydrogenation of flavylium salts is generally simple and results in the saturation of the hetero-ring. There is an interesting and unexplained difference between LAH reduction at C 2 and sodium borohydride reduction at C 4.

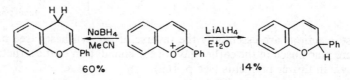

60%   14%

D   Alkyl 1-Benzopyrylium Salts

Alkyl substituents at C 2 and at C 4 behave like the corresponding pyrylium derivatives and, for example, can be condensed with benzaldehyde.

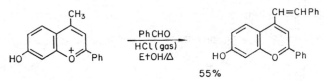

55%

E   Anthocyanidins

These are polyhydroxy flavylium salts; they occur in a large proportion of the red to blue flower pigments. Anthocyanidins are generally bound glycosidically to sugar residues, and these glycosides are known as anthocyanins: as an example, cyanin chloride is an anthocyanin and occurs in the petals of the red rose (*Rosa gallica*), the poppy (*Papaver rhoeas*), and very many other flowers; another example is malvin chloride, which has been isolated from many species, among which is *Primula viscosa*, a mauvy-red alpine primula.

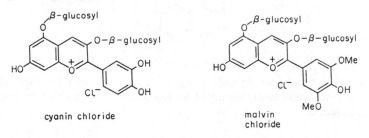

cyanin chloride

malvin
chloride

In the living cell these compounds probably exist in much more complex bound forms, and the colour any one pigment shows can vary with the precise cellular environment in which it finds itself. Though the actual colours *in vivo* are not simply controlled by pH changes, it is nevertheless instructive to see how the colour of a typical anthocyanidin, cyanidin, is strongly pH-dependent.

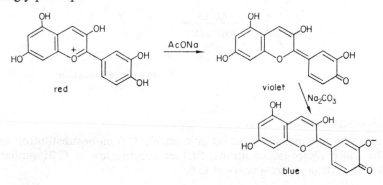

red

violet

blue

Each of the above structures is only one of a number of possible canonical forms, and the deep colours of these compounds are the consequence of extensive mesomeric delocalization of charge, e.g.:

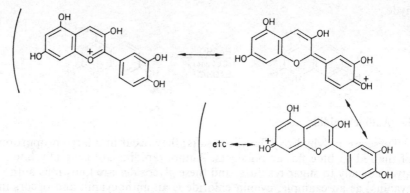

REACTIONS OF COUMARINS

coumarin

Coumarins have a much wider range of reactivity than 1-benzopyrylium salts, they are quite easily attacked by electrophilic, as well as by nucleophilic, reagents.

A Reactions with Electrophilic Reagents

1 ADDITION TO CARBONYL OXYGEN

Coumarins are not appreciably protonated in aqueous strong acids and pK_a values have not been reported. However, coumarin does react with Meerwein's reagent to give 2-ethoxy-1-benzopyrylium borofluoride.

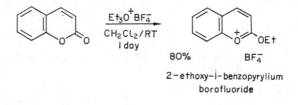

2 SUBSTITUTION AT CARBON

(a) *Nitration and sulphonation* give mainly C 6 monosubstitution and, under more vigorous conditions, further substitution at C 3: similarly, Friedel-Crafts *acylation* occurs at C 6.

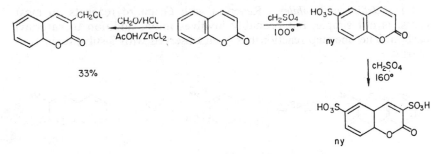

33%

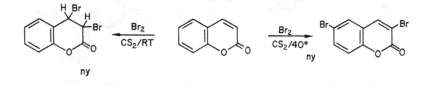

(b) *Chloromethylation* occurs at C 3. It is quite likely that this represents electrophilic attack of coumarin itself, whereas under the much more strongly acidic conditions of sulphonation, nitration, and acylation, it is an O-protonated or O-AlCl₃-complexed coumarin which is reacting.

(c) *Bromination* under mild conditions results in addition.

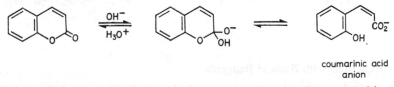

B   Reactions with Oxidizing Agents

Non-phenolic coumarins are relatively stable to oxidation which, when it does occur, leads to general breakdown.

C   Reactions with Nucleophilic Reagents

(a) *Hydroxide ion.* Coumarins are hydrolysed by alkali to salts of coumarinic acids. The free acids, however, which retain the *cis*-geometry of the double bond, cannot be isolated as they recyclize spontaneously:

coumarinic acid
anion

Prolonged alkali treatment causes isomerization to the *trans*-acids, the coumaric acids, which can be isolated.

(b) *Ammonia and amines* do not lead to α-quinolones, even under forcing conditions, the reasons being the same as those given for 1-benzopyrylium salts (see p. 168).

172   HETEROCYCLIC CHEMISTRY   Ch.

(c) Carbon nucleophiles. Reaction with Grignard reagents, in most cases, is not simple, but probably involves initial addition to carbonyl carbon. The one-step reaction has been observed with 3-substituted coumarins.

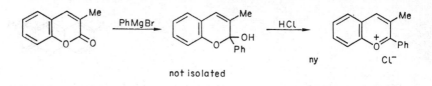

not isolated

Coumarin itself reacts further at C 2 or C 4 with a second molecule of Grignard reagent, possibly by way of a ring-opened intermediate, for example with MeMgI:

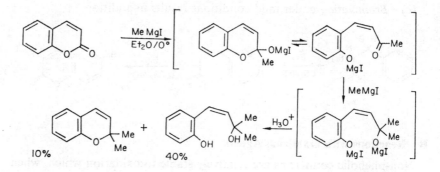

Milder carbon nucleophiles, such as cyanide anion or malononitrile anion which more usually add to the β-carbon of $\alpha\beta$-unsaturated carbonyl compounds, react with coumarin at C 4.

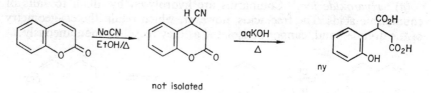

not isolated

D   Reactions with Radical Reagents

Very little work, other than hydroxylation of the benzene ring, has been done in this area.

E   Reactions with Reducing Agents

Catalytic and LAH reductions proceed quite normally.

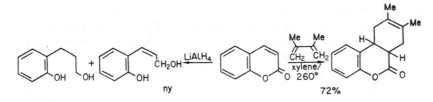

F Reactions with Dienes

Coumarin can act as a dienophile in Diels-Alder reactions (cf. α-pyrone, p. 158).

REACTIONS OF CHROMONES AND FLAVONES

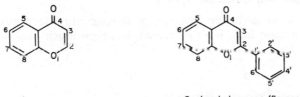

chromone   2-phenyl chromone (flavone)

Chromones and flavones react easily with electrophilic and nucleophilic reagents, as do the coumarins. Even though an enormous amount of work has been done on the phenolic chromones and flavones, most of it has been of a degradative character, and much fundamental work remains to be done with simple chromones.

A Reactions with Electrophilic Reagents

1 ADDITION TO CARBONYL OXYGEN

Chromones are much more basic than coumarins, thus a mixture of a coumarin and a chromone can be separated by treatment of an ether solution of the two with dry hydrogen chloride: the 4-hydroxychromylium chloride precipitates out, and the coumarin remains in solution. The pK_a of chromone itself is 2·0, and that of flavone is 1·3.

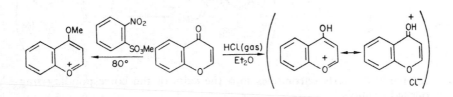

Methylation of carbonyl oxygen has been achieved with strong methylating agents such as methyl o-nitrobenzenesulphonate.

2   SUBSTITUTION AT CARBON

Substitution occurs in the benzene ring at C 6 or C 8 under strongly acidic conditions, and at C 3 under more weakly acidic conditions, just as in the case of coumarin.

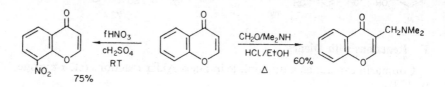

Bromination under mild conditions gives addition to C 2-C 3.

B   Reactions with Oxidizing Agents

Ozonolytic or peroxidic cleavage of flavones, in which any phenolic hydroxyls present have been protected, has been much used in structure determination.

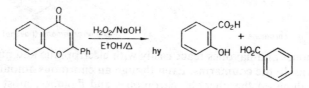

C   Reactions with Nucleophilic Reagents

(a) *Hydroxide and alkoxide ions.* Cold aqueous sodium hydroxide

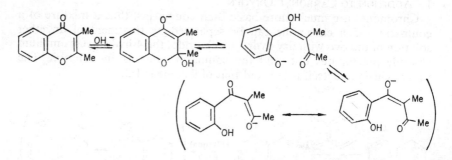

reversibly converts chromones into the salts of the corresponding ring-opened phenols.

More vigorous treatment causes fission of the 1,3-diketone side-chain.

Under different conditions, base catalysis leads to the formation of a dimeric product, 4: the essential step here is likely to be nucleophilic

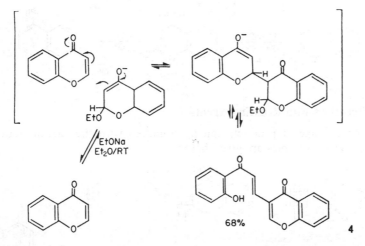

addition of the enolate anion (formed by OEt⁻ addition to chromone) to the C 2 of chromone itself.

(b) *Ammonia and amines.* Addition to C 2 is followed by ring-opening e.g.:

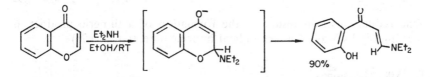

Recylization to γ-quinolones does not occur (cf. p. 168).

Chromones form normal carbonyl derivatives only in a limited number of cases: thus, flavone reacts with tosylhydrazine, but not with phenylhy-drazine, to give a 'normal' derivative, 5, by attack on carbonyl carbon.

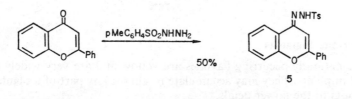

(c) *Reactions with carbon nucleophiles.* Grignards react with chro-mones at carbonyl carbon and the resulting chromenols are converted by acid into the corresponding 1-benzopyrylium salts.

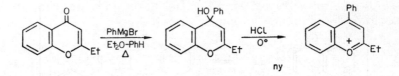

D   Reactions with Reducing Agents

A wide range of products can be obtained with the various reducing agents. Two examples are given below:

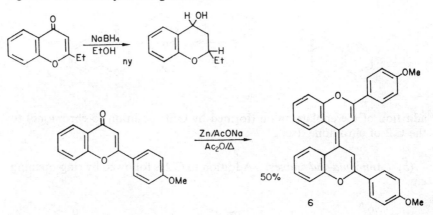

Zinc reduction again results in the formation of a dimeric product, 6, presumably by way of a free radical intermediate. (c.f. p. 153)

E   Alkylchromones

Alkyl groups at C 2, but not at C 3, are activated and can be condensed with carbonyl compounds or oxidized by selenium dioxide.

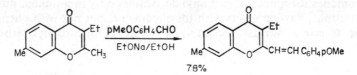

F   Flavone Pigments

The naturally-occurring flavones are yellow and are very widely distributed in plants. They may accumulate in almost any part of a plant, from the roots to the flower petals.

Unlike the anthocyanins, which are too reactive and short-lived, the very much stabler flavones have, from time immemorial, been used as dyes, for they impart various shades of yellow to wool. As an example, in the more recent past the inner bark of one of the North American oaks,

Quercus velutina, was a commercial material known as quercitron bark and much used in dyeing: it contains quercitrin.

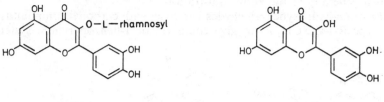

quercitrin   quercetin

The corresponding aglycone, quercetin, is one of the most widely-occurring flavones, found, for example, in *Chrysanthemum* and *Rhododendron* species, horse chestnuts, lemons, onions, and hops.

Another example of a flavone glycoside is apiin, which has been isolated from celery and parsley.

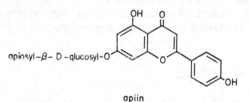

apiin

THE SYNTHESIS OF 1-BENZOPYRYLIUM SALTS, CHROMONES AND COUMARINS

The most important general ways of putting together these heterocyclic systems can be divided into three categories. All three begin with benzene compounds. The general similarity with quinoline syntheses may be noted.

(*a*) Subject to the restrictions set out in the detailed discussions which follow, phenols can react with 1,3-dicarbonyl compounds to produce

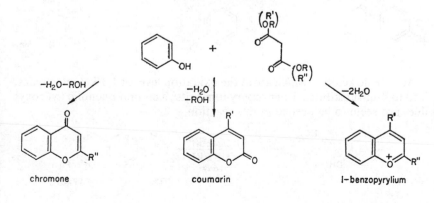

chromone   coumarin   1-benzopyrylium

1-benzopyrylium, coumarin or chromone systems, depending on the oxidation level and oxygenation pattern of the starting materials. This is shown generally in the above illustration.

(b) *ortho*-hydroxybenzaldehydes can react with carbonyl compounds of the type R—CO.CH$_2$.R' to give coumarin or 1-benzopyrylium systems,

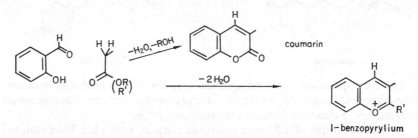

depending on the nature of the aliphatic unit.

(c) *ortho*-hydroxyacetophenones can react with esters (or aldehydes, followed by an oxidation step) to give chromones.

A   Ring Synthesis of the 1-Benzopyrylium System

(a) *From phenols and 1,3-diketones or 1,3-ketoaldehydes or their equivalents.*   The simplest reaction, that between a diketone and a phenol, works best with resorcinol, for the second hydroxyl facilitates the cyclizing electrophilic attack on the benzene ring: since this reaction gives mixtures with unsymmetrical diketones, it is best suited to the synthesis of symmetrically substituted 1-benzopyrylium compounds.

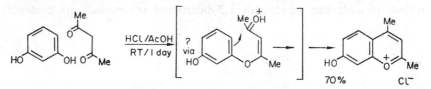

Acetylenic ketones, which are at the oxidation level of 1,3-ketoaldehydes, lead to 4-unsubstituted 1-benzopyrylium salts; a second phenolic hydroxyl does not seem to be needed in this reaction.

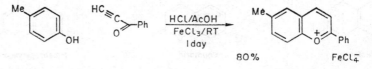

(b) *From o-hydroxybenzaldehydes and ketones.* Salicylaldehydes can be condensed by acid or base catalysis with ketones of the type R—CO.CH$_2$.R'. When base catalysis is used, the intermediate hydroxy chalcones can be isolated, but overall yields are often better when the whole sequence can be carried out in one step using acid catalysis.

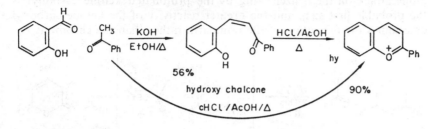

56%

hydroxy chalcone

cHCl /AcOH/Δ

B   Ring Synthesis of Coumarin Systems

(a) *From o-hydroxybenzaldehydes and esters or carboxylic anhydrides.* The simplest synthesis of coumarins is a special case of the Perkin condensation. The o-hydroxy-*trans*-cinnamic acids cannot be intermediate since they do not isomerize under the conditions of the reaction; nor can O-acetylsalicylaldehyde be the immediate precursor of the coumarin, since it is not cyclized by sodium acetate on its own. A probable reaction sequence is shown:

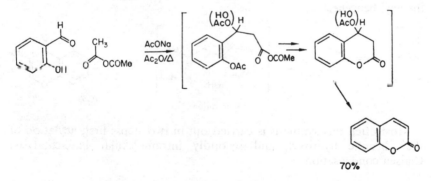

70%

A second synthesis, which also starts from o-hydroxybenzaldehydes, involves malonic ester, and proceeds under mild conditions: since coumarin 3-carboxylic acids are easily decarboxylated, this provides a route to 3-unsubstituted coumarins. β-Keto esters under these conditions give 3-acylcoumarins.

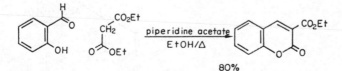

80%

(b) *From phenols and 1,3-ketoesters.* Phenols react with β-ketoesters to give coumarins under most conditions. In principle they could react in the alternative orientation to give chromones, but this can only be achieved under rather special conditions.

This approach works best with the more reactive resorcinols: electrophilic attack on the benzene ring by the protonated ketone carbonyl is the probable first step, and the greater reactivity of the ketone carbonyl determines the formation of the coumarin rather than the chromone.

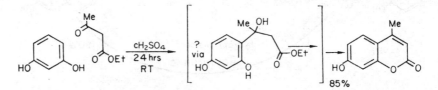

Phenol, being less nucleophilic, requires a more powerful acidic catalyst, so that, for example, hydrofluoric acid at 100° has been used with acetoacetic ester to give a 60 per cent yield of 4-methylcoumarin.

C   Ring Synthesis of Chromone Systems

(a) *From o-hydroxyacyl benzenes and esters.* This synthesis involves a Claisen condensation between an ester and the activated methylene of the acyl benzene.

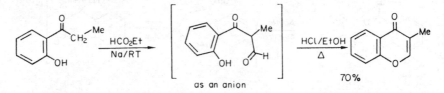

Most often this synthesis is carried out in two steps: first, acylation of the phenolic hydroxyl, and secondly intramolecular base-catalysed Claisen condensation.

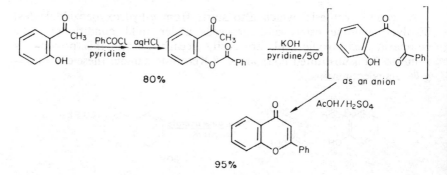

(b) *From o-hydroxyacyl benzenes and an aromatic aldehyde.* Base-catalysed aldol condensation gives a chalcone, which may be cyclized to a flavanone which is then dehydrogenated in a variety of ways to a flavone. The example illustrates the use of triphenylmethyl perchlorate as a hydride abstractor in the dehydrogenation:

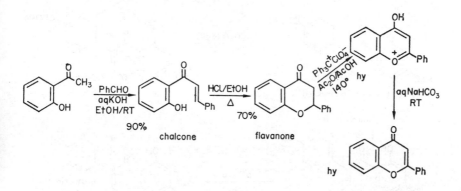

D   Examples of 1-Benzopyrylium, Chromone and Coumarin Syntheses

(a) *Pelargonidin chloride*, from, for example, antirrhinums.

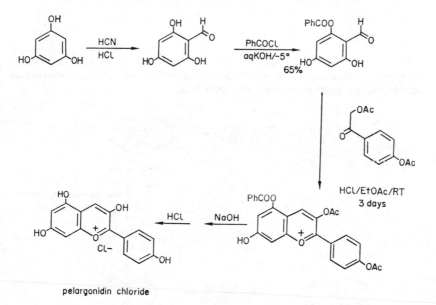

pelargonidin chloride

(b) *Prunetin*, from the bark of wild cherry.

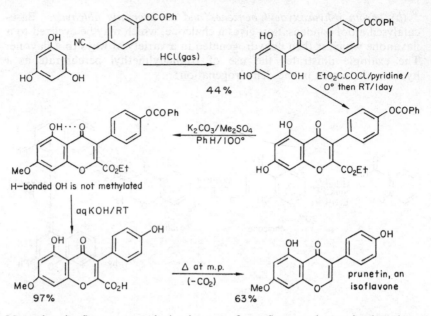

H-bonded OH is not methylated

Note that isoflavones are derived *in vivo* from flavones by aryl migration.

(c) *Umbelliferone*, a highly fluorescent constituent of *Daphne mezer-eum* and angelicine, from *Angelica archangelica;* in Java derivatives of angelicine are used as poisons to catch fish.

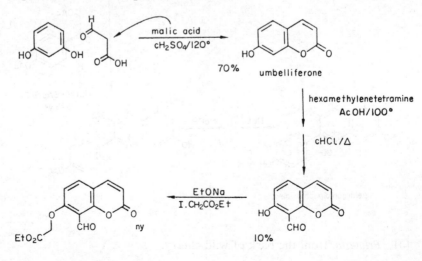

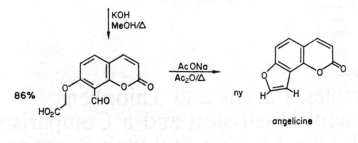

15

Pyrroles, Furans and Thiophens: General Discussion and a Comparison with One Another and with Benzene Compounds

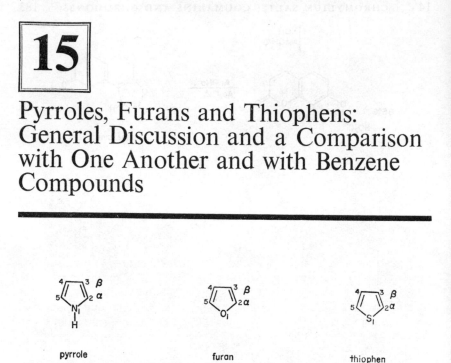

pyrrole   furan   thiophen

Electrophilic Attack at Carbon

These three heterocycles are characterized mainly by a very high degree of reactivity towards substitution by electrophilic reagents. This is easy to understand in that, as was noted when we considered their structures, they contain six π-electrons distributed over five ring atoms: they are what is known as electron-rich, or, as stated by some authors, 'π-excessive'.

Mesomeric electron-release from the heteroatoms facilitates attack by electrophiles at the ring carbons, and this characteristic reaction finds a parallel in the reactions of enamines, enol ethers, and thioenol ethers, and also in the reactions of anilines, phenol ethers, and thiophenol ethers.

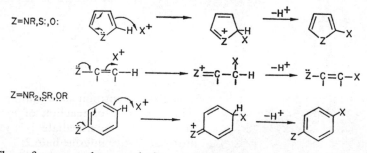

Thus, for example, acetylation and allylation of cyclohexanone pyrrolidine enamine can be seen to be analogous with the corresponding reactions of pyrrole.

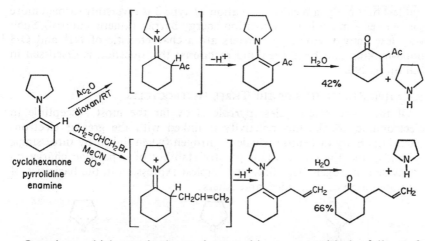

cyclohexanone
pyrrolidine
enamine

Questions which puzzle the student at this stage are (*a*) the failure of, say, methyl iodide to react with the nitrogen of N-methylpyrrole; (*b*) predominant C-methylation of cyclohexanone pyrrolidine enamine with methyl iodide whereas dimethylaniline or trimethylamine react easily and rapidly on nitrogen to give quaternary ammonium salts; and (*c*) the failure of acetic anhydride to acetylate pyrrole on nitrogen.

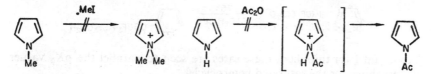

These two questions are related. The simple argument used to explain this situation is as follows. Assume the transition state of the addition of the electrophile to the heterocycle resembles that of the cationic intermediate, then since the positive charge of the product of addition of the electrophile to carbon is stabilized by delocalization, whereas the positive charge in the product of addition to nitrogen is entirely localized on the nitrogen, the latter type of cation, 2, is much less stable than the former, 1. Therefore,

the activation energy for the formation of the transition state leading to 2 is higher than that leading to 1. Thus, the rate of reaction of methyl iodide or of acetic anhydride at carbon to give intermediate 1 is much higher than the rate of reaction at nitrogen to give intermediate 2.

The general situation, however, is much more complex, for this argument does not hold for all electrophilic additions: protonation is the extreme example where, even though the cation of type 1 is still the stablest thermodynamically, the cation of type 2 is reversibly formed more easily over a much lower activation energy barrier (kinetic control). Such very low activation energy barriers are a characteristic of NH and OH bond-making and bond-breaking reactions: this question is discussed in full on p. 290.

RELATIVE REACTIVITIES OF THE THREE HETEROCYCLES

Of the three heterocycles, pyrrole is by far the most susceptible to electrophilic attack: this reactivity is linked with the greater electron-releasing ability of neutral trivalent nitrogen (when linked by three single σ bonds) and the concomitant greater stability of a positive charge on tetravalent nitrogen. This finds its simplest expression in the basicities of saturated amines, sulphides, and ethers:

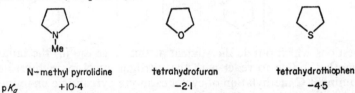

	N–methyl pyrrolidine	tetrahydrofuran	tetrahydrothiophen
pK_a	+10·4	−2·1	−4·5

The relative rates of trifluoroacetylation of pyrrole, furan, and thiophen by trifluoroacetic anhydride are $5·3 \times 10^7$ for pyrrole, $1·4 \times 10^2$ for

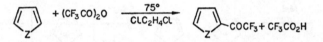

furan, and 1 for thiophen: these rates are seen to parallel the pK_a values quoted above for the saturated heterocycles.

The greater reactivity of pyrrole is also seen clearly in its rapid reaction with weak electrophiles such as PhN$_2^+$ and HNO$_2$, neither of which reacts with furan or thiophen. It is relevant to note here that dimethylaniline reacts rapidly with PhN$_2^+$ and HNO$_2$, but anisole does not, and also that enamines are more reactive than enol ethers.

COMPARISON WITH BENZENE ITSELF

Benzene is, as might be expected, much less reactive than thiophen towards electrophilic reagents: the probably greater resonance stabilization of benzene must be in part responsible, but another important factor is the higher energy of the transition state which is structurally related to the intermediate benzenonium cation, in which the positive charge resides only on carbon.

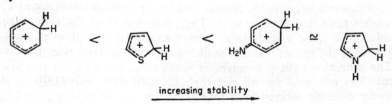

increasing stability

Rate differences between systems, however, depend very much on the nature of the reagent: thus, whereas thiophen is protonated 10^3 times more rapidly than benzene in aqueous sulphuric acid, it is brominated by molecular bromine in acetic acid about 10^9 times more rapidly than benzene. In comparing the reactivities of different heterocycles, one must be very careful not to use different reactions.

RELATIVE REACTIVITIES OF α- AND β-POSITIONS

This leads us to consider why electrophilic attack occurs more readily at the α- than at the β-position. This is most easily understood in terms of resonance stabilization or delocalization of positive charge in the intermediate cation, which in the structures below is seen to be greater in the

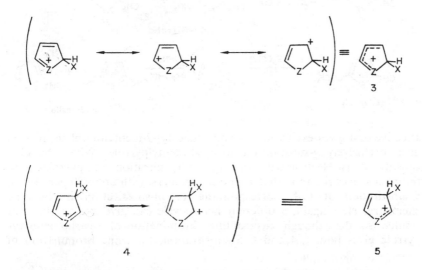

case of the cation 3, derived by α-addition. Note carefully that in the cation 4, derived by β-addition, the C 4-C 5 double bond is not and cannot be mesomerically involved in the delocalization of the positive charge. This difference between α- and β-reactivity is most marked in furan and least marked in pyrrole.

Variation of rate differences. Just as rate differences between heterocycles vary with reagent, so rate differences between reaction at an α- or β-position vary with reagent. For example, nitration of thiophen gives a 6 : 1 ratio of 2- to 3-nitrothiophen, whereas monochlorination appears to give exclusively 2-chlorothiophen. This is an aspect of the general rule that the more reactive a reagent is, the less selective it is: in other words, the reagent can be so reactive that it will not differentiate very markedly between the various available ring-positions.

Free radical substitutions at this stage have not been sufficiently studied to allow generalizations to be made.

ELECTROPHILIC SUBSTITUTION OF 2-MONOSUBSTITUTED COMPOUNDS

Furans behave quite simply in that, whatever the electrophile, and irrespective of whether the C 2 substituent is electron-releasing or electron-withdrawing, the dominant point of attack is always C 5, the other α-position.

By contrast, C 2-substituted thiophens and pyrroles can give isomer mixtures: when the C 2-substituent is electron-releasing, then nitration gives the 2,5-isomer as major product and the 2,3-isomer as minor product, which corresponds with *p*- and *o*-substitution in benzene chemistry. A milder electrophile, uncatalysed trifluoroacetic anhydride, is much more

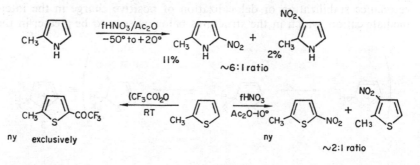

selective and gives exclusively 5-trifluoroacetyl-2-methylthiophen, it even gives exclusively α-substitution with N-methylpyrrole. When the C 2-substituent is electron-withdrawing, then nitration of pyrroles and thiophens gives rise to both 2,4- and 2,5-isomers, with up to 80 per cent of β-substitution at C 4, corresponding to *m*-substitution in benzene chemistry. Here again, a different heterocycle can give rise to different results, so that though electrophilic bromination of 2-carbomethoxy-pyrrole gives both C 4 and C 5 substitution, the same bromination of

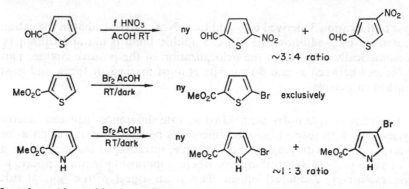

2-carbomethoxythiophen gives exclusive C 5-substitution.

The reason why α-substitution should be so much more strongly favoured over β-substitution in furans than in thiophens and pyrroles is not clearly understood, but may be connected with the tendency of furans to form intermediates by 2,5-addition, which in turn can be related to their lower reactivity towards electrophiles (see p. 293), and their lower aromaticity.

ELECTROPHILIC SUBSTITUTION OF 2,5-DISUBSTITUTED COMPOUNDS

In such compounds, simple substitution of β-hydrogen occurs quite readily in most cases. In some, however, electrophilic displacement of one of the α-substituents is observed. This occurs most frequently with α-carboxylic acids (see pp. 210–242), but also occurs with, for example, halogen compounds (p. 225) and ketones (p. 247). This type of displacement is a simple consequence of the fact that electrophilic *addition* can occur to any reactive position whether it is substituted by hydrogen or by other groups. In the majority of instances the electrophile which has added ionizes off again, so that no change is observed; in some instances, however, where the activation energy for the alternative ionization is appropriate, the group originally present as a substituent ionizes off.

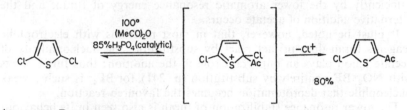

ADDITION REACTIONS

An important feature of furan chemistry is the occurrence of a number of 2,5-addition reactions, the best-known of which is the addition of acetyl nitrate (p. 241); thiophen shows a very much smaller tendency to add, as is seen in the isolated example of the formation of a small proportion of tetrachlorotetrahydrothiophen in reaction with chlorine (see p. 222); pyrrole does not react in this manner.

Such reactions begin by electrophilic addition to C 2 to give the usual intermediate cation, 3 which is followed by nucleophilic addition to C 5 of whatever is the most reactive nucleophile present.

Let us look at the reaction of furan and pyrrole with acetyl nitrate:

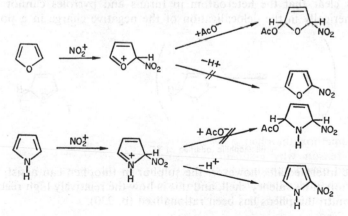

In both cases the initial reaction is addition of NO_2^+ to give an intermediate cation: what happens next is determined by the relative activation energies, of deprotonation to give an aromatic compound and of simple nucleophilic addition to give a non-aromatic compound. If one makes the reasonable assumption that the aromatic system is partially developed in the transition state of the deprotonation reaction, then the higher the

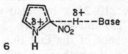

degree of aromatic stabilization of the product, the lower the activation energy for the deprotonation: hence the pyrrolium cation intermediate is deprotonated *via* 6 because of the greater aromatic resonance energy of the pyrrole ring system. In the furan case deprotonation just fails to be assisted sufficiently by the lower aromatic resonance energy of furan, and the alternative addition of acetate occurs.

It must be noted, however, that in most reactions with electrophilic reagents furan does in fact react by substitution, the nucleophilicity of acetate anion plays an important role in the addition: thus furan reacts with $NO_2^+BF_4^-$ entirely by substitution (p. 241), for BF_4^- is such a weak nucleophile that deprotonation becomes the favoured reaction.

The lower resonance stabilization of furan is also seen in its behaviour as a 1,3-diene in Diels-Alder type reactions (p. 245): this type of reaction is very rare in pyrrole (p. 206) and in thiophen chemistry (p. 228).

Reaction with Nucleophilic Reagents

Nucleophilic substitution of H is completely unknown in pyrrole, furan, and thiophen chemistry, and this is paralleled by the chemistry of simple benzene hydrocarbons; substitution of halogen substituents does occur, but vigorous conditions, rather similar to those necessary to effect displacement of halide in halobenzenes, have to be used.

It is clear that the heteroatom in furans and pyrroles cannot assist mesomerically in the delocalization of the negative charge in a possible

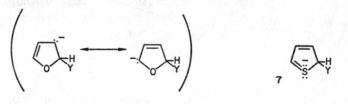

nucleophilic addition of Y⁻

anionic intermediate: however, the sulphur in thiophen can assist, 7, by expansion of its valency shell, and this is how the relatively high reactivity of halonitrothiophens has been rationalized (p. 230).

The as yet not properly understood H-D exchange in furan-, thiophen-, and N-methylpyrrole-2-carboxylic acids in 40 per cent $NaOD-D_2O$ must be mentioned in this context. This reaction provides the best route to tetradeuteriofuran.

Oxy-Derivatives

In all three heterocycles, an oxy substituent exists mainly in the carbonyl tautomeric form because of the usual mesomeric stabilization in which the heteroatom acquires a partially positive charge.

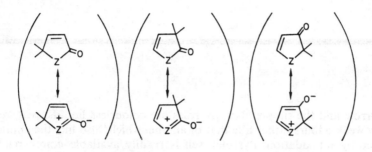

Carbonyl Derivatives

Mesomeric interaction between carbonyl and the ring systems also involves partial positive charges on the heteroatom. Such interaction

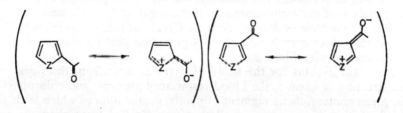

reduces the reactivity of the carbonyl group towards nucleophilic reagents: the effect is of course most marked in the pyrrole series, where, for example, although N-methylpyrrole-2-aldehyde reacts with the usual carbonyl reagents, it does not undergo the Cannizzaro or Perkin reactions; furfural (furan-2-aldehyde) and thiophen-2-aldehyde, however, do undergo these reactions.

Pyrroles: Reactions and Synthesis

Pyrrole and the simple alkyl pyrroles are colourless liquids, with relatively weak odours rather like that of aniline, which, also like the anilines, darken by autoxidation. Pyrrole itself is readily available commercially, and is manufactured by alumina-catalysed gas-phase interaction of furan and ammonia.

Pyrrole was first isolated in 1857 from the pyrolysate of bone: it probably arises in this rather extraordinary process by a sequence similar to that which was until recently the main method for the preparation of pyrrole, the pyrolysis of the ammonium salt of the sugar acid, mucic acid.

The word pyrrole is derived from the Greek for red, which refers to the bright red colour pyrrole imparts to a pinewood shaving moistened with concentrated hydrochloric acid.

The main impetus for the study of pyrroles came from the work on the structure of haemin, the blood respiratory pigment, and chlorophyll, the green photosynthetic pigment of plants, degradation of which leads to the formation of a mixture of alkylpyrroles. In fact these pigments are synthesized in the living cell from porphobilinogen, the *only* aromatic pyrrole to play a function—a vitally important function—in fundamental metabolism.

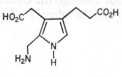

porphobilinogen

Ultimately all life on earth depends on the incorporation of atmospheric carbon dioxide into carbohydrates. The energy for this highly endergonic process is sunlight, and the whole is called photosynthesis. The very first step in photosynthesis is the absorption of a photon by pigments, of which the most important in multicellular plants is chlorophyll-*a*. This photonic energy is then used chemically to achieve the crucial $CO_2 \rightarrow C$ bonding

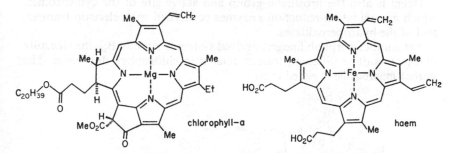

chlorophyll-*a*

haem

reaction, which is a reductive carboxylation in which oxygen is liberated, and can be expressed in the following very simplified way.

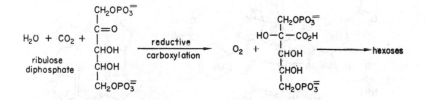

The presence of the by-product of this process, molecular oxygen, allowed the evolution of aerobic organisms of which man is one: haemoglobin is the agent in the blood-stream which carries oxygen from lung to tissue in mammals and it is made up of the protein globin associated with a prosthetic group, the pigment haem (also spelt heme). The very close similarity of this with chlorophyll is striking, and points to a common evolutionary origin.

The iron in haem is six-coordinate Fe(II) which, in oxygenated haemoglobin, is liganded below the plane of the macrocycle with an imidazolyl

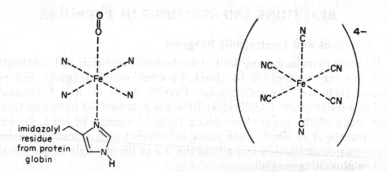

imidazolyl residue from protein globin

residue from the protein, and above the plane with molecular oxygen. Ferrocyanide anion has iron in the same state of coordination.

The metal-free, unsubstituted tetrapyrrolic macrocycle is called porphyrin, and the haem without the ferrous ion is called protoporphyrin IX.

Haem is also the prosthetic group and active site of the cytochromes, which are oxidation-reduction enzymes concerned with electron transfer, and of the hydroperoxidases.

Yet another porphobilinogen-derived system is vitamin B_{12}, the structure of which is quite different, though related to chlorophyll and haem. The parent macrocycle is called corrin.

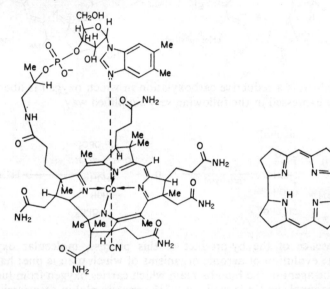

Vitamin B_{12} corrin ring system

Many secondary metabolites contain the aromatic pyrrole unit, and some of these are useful antibiotics; however, not many synthetic chemotherapeutic compounds contain it.

REACTIONS AND SYNTHESIS OF PYRROLES

A Reactions with Electrophilic Reagents

Whereas pyrroles are resistant to nucleophilic addition and substitution, they are very susceptible to attack by electrophilic reagents and react almost exclusively by substitution. Pyrrole itself, N- and C-monoalkyl and to a lesser extent C-dialkylpyrroles, are polymerized by strong acids so that many of the usual electrophilic reagents cannot be used. However, the presence of an electron-attracting substituent such as methoxycarbonyl prevents polymerization and allows the use of the strongly acidic nitrating and sulphonating reagents.

(a) Protonation. Reversible proton addition occurs at all positions, being by far the fastest at the nitrogen, and about twice as fast at C 2 as at C 3. Thermodynamically the stablest cation, the 2H-pyrrolium ion, is that formed by protonation of C 2 and the observed pK_a values refer to this species (see p. 290 for a discussion of N versus C protonation). The weak N-basicity of pyrroles is the consequence of the absence of mesomeric delocalization of charge in 1H-pyrrolium cation.

The pK_a values of a wide range of pyrroles have been determined: pyrrole itself is an extremely weak base with a pK_a of $-3\cdot8$; this, as a $0\cdot1$ molar solution in normal acid, corresponds to one protonated molecule to about 5000 unprotonated. Basicity increases very rapidly with increasing alkyl substitution, so that 2,3,4,5-tetramethylpyrrole, with a pK_a of $+3\cdot7$, is almost completely protonated on carbon as a $0\cdot1$ molar

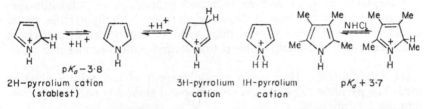

2H-pyrrolium cation (stablest)   3H-pyrrolium cation   1H-pyrrolium cation $pK_a +3\cdot7$

solution in normal acid (cf. aniline, which has a pK_a of $+4\cdot6$). This illustrates in a most striking manner the stabilizing effect which methyl groups have on cations.

Reactions of Protonated Pyrroles. The 2H- and 3H-pyrrolium cations are essentially immonium ions and as such are electrophilic: they play the key role in polymerization (see p. 201) and in reduction (p. 205). Two other reactions in which they are involved are noteworthy: the first of these

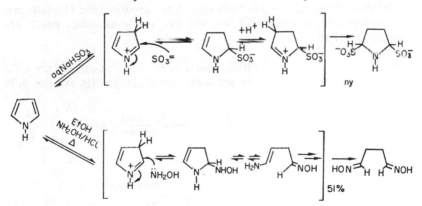

is addition of sodium bisulphite; the second is reaction with hydroxylamine which opens the pyrrole ring to give succindialdehyde dioxime. The more reactive 3H-pyrrolium cation is probably the starter in both cases and the schemes shown represent likely reaction paths.

(*b*) *Nitration.* Nitrating mixtures suitable for benzenoid cause complete decomposition of pyrrole, but reactio smoothly with acetyl nitrate in acetic anhydride at low temperature. The reagent is formed by mixing fuming nitric acid with acetic anhydride (care!) to give acetyl nitrate and acetic acid, the strong mineral acid disappears completely. In nitration of pyrrole with this reagent it has been shown that C 2 is $1·3 \times 10^5$ and C 3 is 3×10^4 times as reactive as benzene.

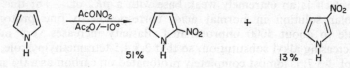

N-alkylated pyrroles give an increased proportion of β-substitution, 2 : 1 of α to β in the case of N-methyl pyrrole. 2-Methylpyrrole, under comparable conditions, gives a poor yield of mononitro derivatives in which the ratio of 2-methyl-5-nitro to 2-methyl-3-nitro-isomer is 6 : 1.

(c) *Sulphonation.* Here again a mild reagent of low acidity must be used. The pyridine sulphur trioxide compound smoothly converts pyrrole into the 2-sulphonate.

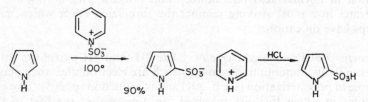

β-Sulphonation occurs when both α-positions are blocked. The salts can be converted with mineral acid into the free sulphonic acids which are relatively stable strong acids.

(d) *Halogenation.* Pyrrole halogenates so readily that unless very mild conditions and special reagents are used, tetrahalopyrroles are the only

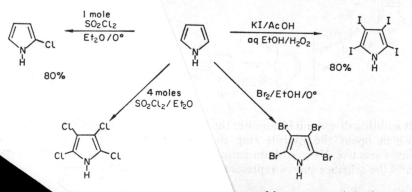

ducts. 2-Chloropyrrole, a very unstable compound, is the only ...asily prepared by direct halogenation. Attempts to monohalogenate simple alkylpyrroles have failed, probably because of benzylic halogenation leading to the extremely reactive pyrryl alkyl halides.

(e) *Acylation.* Direct acetylation of pyrrole with acetic anhydride at 150–200° leads to 2-acetyl and 2,5-diacetylpyrroles, and no N-acetylpyrrole: a small proportion of the latter is formed, however, in the presence of sodium acetate. It is noteworthy that no reaction occurs below 100°. Increasing alkyl substitution facilitates acylation, so that 2,3,4-trimethylpyrrole yields the 5-acetyl derivative even on refluxing in acetic acid.

Useful acylation procedures involve the pyrryl Grignard reagent (p. 208) and the Gattermann or Hoesch reactions, in which protonated hydrogen cyanide or a protonated nitrile is the electrophile and which, because of the strongly acidic conditions, can only be used with di- and tri-alkyl pyrroles.

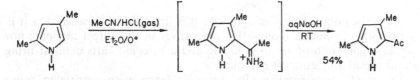

Related to this but much more generally applicable is the Vilsmeier reaction, which provides the most effective way of introducing a formyl

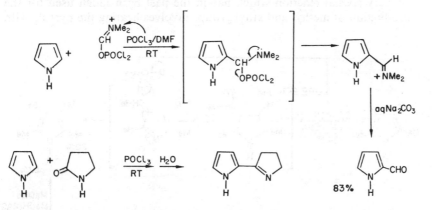

group. It can be used with pyrrole itself, as the reagent does not cause polymerization.

N-acylation is usually achieved by the interaction of pyrrylpotassium and an acyl halide (see p. 208). N-acetyl imidazole (p. 306) efficiently acetylates pyrrole on nitrogen at high temperature:

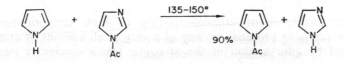

this extraordinary reaction deserves closer study, and is an illustration of the special role of the imidazole ring system (see p. 305).

(*f*) *Alkylation.* Simple pyrroles do not react with methyl iodide below 100°. Above about 150° a series of reactions occurs leading to a complex mixture made up mostly of polymeric material together with some poly-alkylated pyrroles.

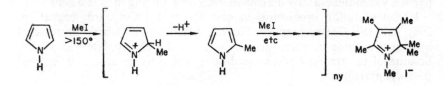

From the point of view of synthesis it is important to realize that it is not possible to obtain mono-alkyl derivatives in this direct manner; nor can a combination of an alkyl halide and a Friedel-Crafts catalyst bring about useful alkylation of pyrroles.

The much more reactive allyl bromide reacts with pyrrole at room temperature in aqueous AcOH-NaOAc to yield a mixture of mono- to tetra-allylpyrroles and dimers, trimers and polymers.

A very special reaction which has in the past been much used for the introduction of methyl and ethyl groups involves heating the pyrrole with

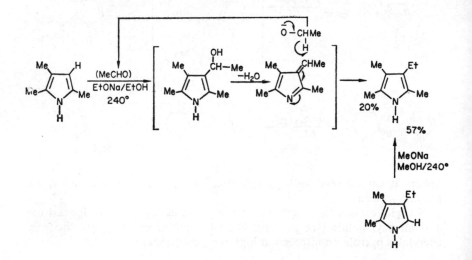

methanol-methoxide or ethanol-ethoxide to high temperatures. The reaction probably proceeds by way of a very small amount of aldehyde, produced by autoxidation of the alkoxide, which condenses (see next section) with the pyrrole and sets up a cycle as shown, aldehyde being regenerated by hydride transfer from alkoxide to alkylidene pyrrolenine.

(g) *Condensation with aldehydes and ketones* occurs easily by acid catalysis (protonation of the carbonyl oxygen) but the resulting pyrrolyl-carbinols cannot be isolated, for under the reaction conditions proton-catalysed loss of water leads to α-alkylidene pyrrolium cations which are highly reactive electrophiles. Thus, in the case of pyrrole itself reaction with aliphatic aldehydes inevitably gives resins, probably linear polymers.

Acetone, however, in reacting in a comparable way, gives a high yield of a cyclic tetramer, 1, maybe because the two methyl groups tend to

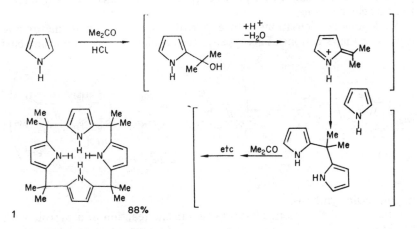

force the pyrrole rings into coplanar conformation (minimum interaction with N-hydrogen or β-hydrogen) thus greatly increasing the chances of cyclization of a linear tetrapyrrolic precursor.

Only when the ring contains an electron-withdrawing group and has only one free α-position (or a free β-position) can high yields of dipyrromethanes be obtained.

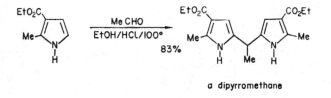

a dipyrromethane

By using an aromatic aldehyde carrying an appropriate electron-releasing substituent the intermediate cation can be sufficiently stabilized by mesomerism to be isolated: such mesomeric cations, 2, are coloured.

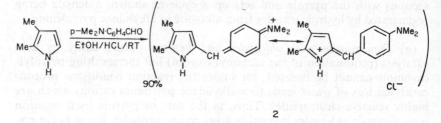

2

The reaction with *para*-dimethylaminobenzaldehyde is the basis of the classical Ehrlich colour reaction: this leads to deep colours not only from pyrroles, but also from furans, indoles (p. 265), etc. which have a free reactive nuclear position.

Analogous condensations with a pyrrole aldehyde lead to mesomeric dipyrromethene cations. This type of cation plays a very important part

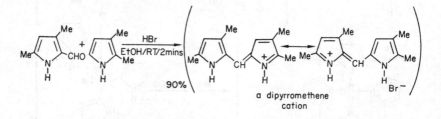

a dipyrromethene
cation

in porphyrin synthesis.

Reduction of the intermediates occurs in the reaction of a pyrrole with an aldehyde or ketone in the presence of hydrogen iodide and hypophosphorus acid, and high yields of product pyrroles are formed in which all

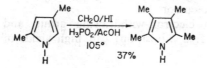

free positions are alkylated. This very useful reaction in synthesis has been used mostly with pyrroles carrying acyl and alkoxycarbonyl substituents, and the substituents are unaffected.

Under alkaline conditions, formaldehyde has been condensed in a

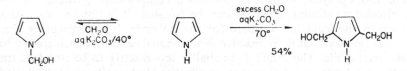

simple manner with pyrrole.

(*h*) *Condensation with imines and immonium ions.* The imine and immonium functional groupings are, of course, the nitrogen equivalents of carbonyl and O-protonated carbonyl groups, and their reactivity is analogous. One of the simplest cases is the reaction of pyrrole with 1-pyrroline, which apparently proceeds by electrophilic attack by neutral C=N.

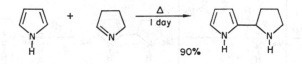

Attack by immonium cation is, as would be expected, very much more rapid. In the Mannich reaction, the immonium cation is formed *in situ* from formalin, dialkylamine and acetic acid. This is a preparatively important reaction.

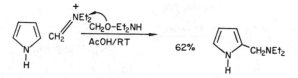

The acid-catalysed polymerization of pyrrole effectively involves a

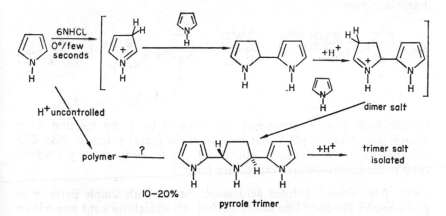

pyrrole trimer

series of Mannich reactions. Pyrrole itself is easily converted by mineral acid into a polymer mixture of unknown composition. However, under controlled conditions, a trimer can be isolated. It is not known if this trimer is the main intermediate for the polymer. It is noteworthy that the less favoured but more reactive β-protonated cation is considered to be the electrophile. The dimer is probably too reactive to be isolated; the trimer salt, however, is not as such an effective electrophile, and is protected from further rapid attack by electrophiles by the positive charge on the central nitrogen atom.

2-Monoalkyl and 2,3-dialkylpyrroles form dimers with ease. Trimers are not formed because the dimer salt is much less electrophilic (because of electron release and steric hindrance by the alkyl) than the protonated monomer and cannot compete with it. Tri- and tetrasubstituted pyrroles do not dimerize.

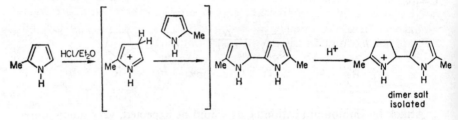

(i) *Carboxylation.* Direct carboxylation of pyrrole occurs readily under pressure, and parallels the carboxylation of phenols.

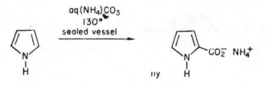

(j) *Diazo-coupling* occurs very readily between pyrroles and benzene diazonium salts.

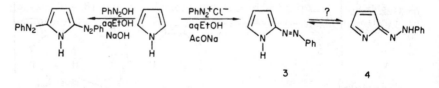

Pyrrole itself gives a mono-azo derivative, 3 or 4, by reacting as a neutral species below pH 8, but by way of the pyrryl anion (p. 204), and 10^8 times faster, in solutions above pH 10. In very strongly alkaline conditions, 2,5-bis-azo derivatives are formed.

(k) *Nitrosation.* Nitrous acid reacts rapidly with simple pyrroles to give complex products (see under oxidation); straightforward nitrosation has only been observed in the presence of an electron-withdrawing group.

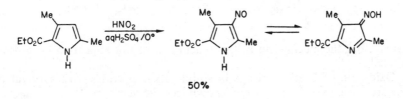

50%

B   Reactions with Oxidizing Agents

Simple pyrroles are generally easily attacked by oxidizing agents, frequently with complete breakdown. When the ring does survive, maleinimide derivatives are the commonest products. This kind of oxidative degradation played an important part in early porphyrin structure determination, in which frequently used oxidizing agents were aqueous sulphuric acid-chromium trioxide or fuming nitric acid.

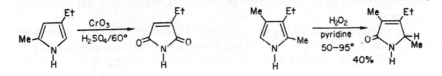

Hydrogen peroxide is a more selective reagent and can convert pyrrole

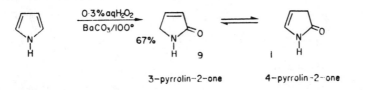

3-pyrrolin-2-one   4-pyrrolin-2-one

itself to a tautomeric mixture of pyrrolin-2-ones in good yield.

Ozonolysis of alkyl pyrroles gives mixtures of dicarbonyl compounds by ring fission. Sensitized photocatalysed reaction of pyrrole with oxygen, which probably proceeds by addition to a cyclic peroxide intermediate, 5,

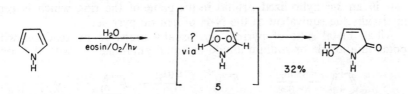

(see p. 226), leads to 5-hydroxy-2-pyrrolone. The rate of autoxidation increases with increasing alkyl substitution. Nitrous acid oxidatively cleaves an α-alkyl group very smoothly to give maleinimide oxime deri-

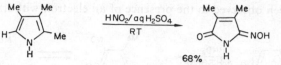

vatives, but the mechanism of this reaction does not appear to have been studied.

Oxidation of C-alkyl groups in simple alkyl pyrroles with survival of the aromatic pyrrole ring is not known (see, however, p. 211).

C   Reactions with Nucleophilic Reagents

Pyrrole and its alkyl derivatives do not react with nucleophilic reagents by addition or by substitution, but only by proton transfer.

1   DEPROTONATION OF N-HYDROGEN

Pyrrole N-hydrogen is very much more acidic (pK_a 17·5) than that of an aliphatic amine, say pyrrolidine (pK_a ~25), or aniline (pK_a ~27) and of the same order as that of 2,4-dinitroaniline (pK_a 15·0). Pyrrole is deprotonated even by hydroxide ion when it is heated with dry potassium hydroxide.

The relatively high acidity of the pyrryl NH can be seen to be mainly the consequence of the mesomerically-derived high fractional positive charge on the nitrogen; another and equivalent way of looking at this is to see that the negative charge of the pyrryl anion is stabilized by mesomeric delocalization, expressed in the following canonical forms: note carefully that, in all these canonical forms, the nitrogen has an unshared electron

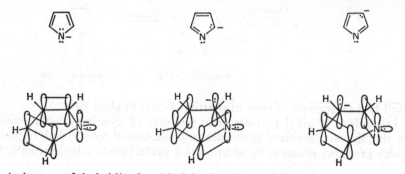

pair in an sp^2 hybridized orbital in the plane of the ring which is geometrically the equivalent of the N-H of neutral pyrrole.

Alkali metal salts of pyrrole are usually prepared by reaction with potassium amide or sodium amide in liquid ammonia, or with the com-

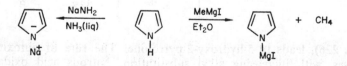

mercially-available *n*-butyllithium in hexane.

Another frequently used reaction is that with a simple Grignard reagent to give the N-Grignard derivative, useful in synthesis (see p. 208). The reactions of these variously metalated pyrroles are dealt with below.

2   DEPROTONATION OF C-HYDROGEN

The α-carbon of N-methylpyrrole is deprotonated first by butyllithium; but, very surprisingly, a second deprotonation to give 2,5- and 2,4-dilithiated derivatives, then occurs. Even using only one mole equivalent of butyllithium, dilithiated species are five times as abundant as monolithiated species. In the presence of excess butyllithium the 2,4-dilithio-derivative is the preferred species and this can be illustrated as shown.

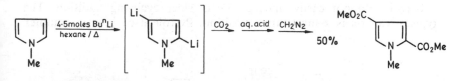

D   Reactions with Radical Reagents

Pyrrole itself tends to give tars under free-radical conditions, probably by way of initial N-hydrogen abstraction. N-Methylpyrrole undergoes free radical benzoyloxylation at α-positions.

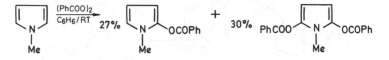

E   Reactions with Reducing Agents

Pyrroles are not reduced by nucleophilic reducing agents such as lithium aluminium hydride, sodium-ethanol or sodium-liquid ammonia, but they are reduced in acidic media, in which the species under attack is very probably the protonated cation. The products are 3-pyrrolines, though in

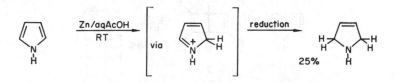

some instances these are accompanied by some of the corresponding 1-pyrroline.

Reduction to pyrrolidines is effected catalytically and occurs much less readily than with furans (p. 244) but seems to be free from complications caused by hydrogenolytic ring-opening. Under milder conditions the pyrrole ring is not reduced. The pyrrole system of 6 provides one of the very

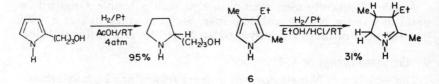

6

few examples of partial catalytic reduction of an aromatic ring (cf. pyrimi-dines, p. 132).

F   Reactions with Dienophiles

Pyrroles do not easily undergo Diels–Alder cyclizing addition. The typical reaction is α-substitutive addition. Pyrrole itself gives a complex

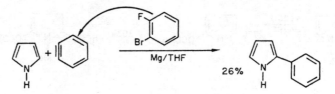

mixture of products with maleic anhydride or maleic acid.

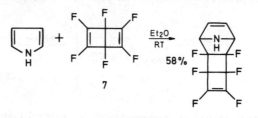

Even benzyne reacts with pyrrole by substitutive addition, however the hexafluoro-bicyclic diene, 7, does react in a Diels–Alder fashion. N-Alkyl, N-aryl, and N-methoxycarbonyl pyrroles show a greater tendency to react by 1,4-cyclo-addition.

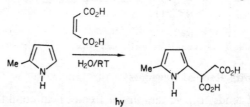

7

G   Reactions with Carbenes

The reaction of pyrrole with dichlorocarbene has been known for a long time, and was at one time one of the main routes to 2-formylpyrrole. The reaction is of particular interest because it in part proceeds by ring

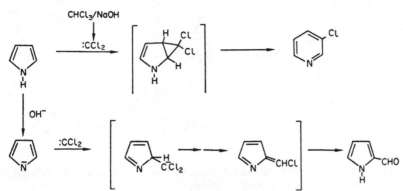

expansion to 3-chloropyridine (this is quite a general reaction, thus indene gives 2-chloronaphthalene).

Chlorocarbene, CHCl, gives a 32 per cent yield of pyridine itself as the only isolable product. Reaction with carbene, CH_2, has not yet been reported.

Carbomethoxycarbene, in contrast with its reaction with furan (p. 246) and thiophen (p. 229), does not give an isolable cyclopropane derivative with N-methylpyrrole. Such a cyclopropane may be an intermediate,

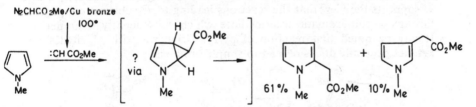

as the scheme suggests, but the isolated product is methyl pyrrylacetate. Electron release by nitrogen may be the main factor in the opening of the three-membered ring intermediate. N-Methoxycarbonylpyrrole, in which the aromatic character is reduced, gives a cyclopropane adduct with carbene.

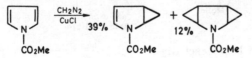

H   Reactions of N-Metallated Pyrroles

The commonly-used lithium, sodium, potassium or magnesium derivatives react very readily with alkyl and acyl halides, with epoxides, and esters. The position of substitution may be the nitrogen, the α-, or the β-carbon or a mixture, depending on the metal and on the solvent used. Broadly speaking the larger the metallic cation and the more polar the solvent, the greater the tendency for N-substitution.

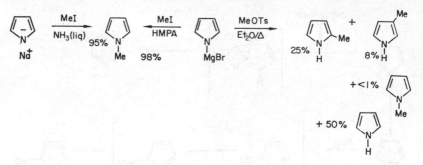

Their reactions with carbonyl halides also bring out the differences caused by varying the metallic cation.

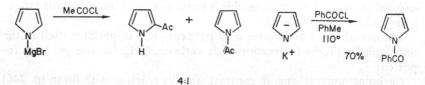

4:1

The fact that 2,3,4,5-tetramethylpyrrole Grignard reacts easily with methyl iodide to give a mixture of the two C-methylated products, 8 and 9, supports the view that the reactions leading to C-substitution proceed through a non-aromatic intermediate which subsequently tautomerizes. It may be noted that reaction of pyrryl Grignards with aldehydes and ketones generally does not lead to simple products.

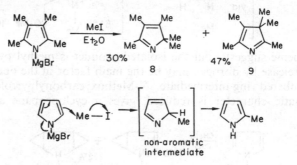

I Alkyl Side-Chain Reactivity

In contrast with the chemistry of pyridine, and even of indole (p. 274). there are no simple examples of alkyl reactivity in simple alkyl pyrroles (see above under halogenation and below under carboxylic esters).

J Pyrryl-C-X Compounds

The pyrrole alcohols of this type, although stable under neutral or basic conditions, are very reactive indeed in acid because of the ease with which they yield quite strongly electrophilic cations. Lithium aluminium hydride reduces such alcohols to alkyl pyrroles. That this proceeds as shown is demonstrated by the failure of lithium aluminium hydride to

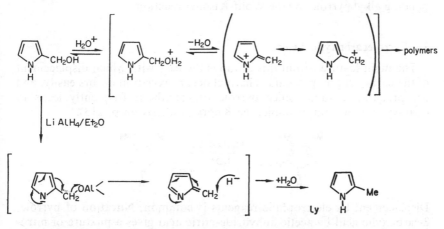

reduce the corresponding N-alkylated compounds under comparable conditions.

The quaternary ammonium compounds typified by 2-dimethylamino-methylpyrrole methiodide (p. 201) are closely related to these alcohols in reactivity and react with nucleophiles by loss of trimethylamine. This has great value in synthesis, as the following sequences show.

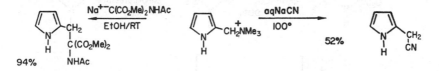

K   Pyrrole Aldehydes and Ketones

These are stable compounds which do not polymerize or autoxidize. The carbonyl groups are less reactive than those in aryl ketones, especially when at the α-position, because of a reduction in the $\delta+$ character of the carbonyl carbon atom by mesomerism involving the ring nitrogen. This is paralleled by the mesomerism in *para*-dimethylaminobenzaldehyde.

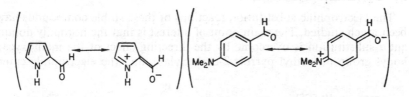

None of these compounds undergoes the Cannizzaro or Perkin reactions, however they do undergo most of the other reactions of ketones and aldehydes. Of particular importance in synthesis are the reduction to the corresponding alcohols by sodium borohydride and to the corresponding alkyl pyrroles by the Wolff-Kishner reaction.

L Pyrrolecarboxylic Acids

The main feature within this group is the ease with which displacement of the carboxyl group occurs. Thermal decarboxylation occurs easily, and has preparative value, since pyrrole ring-synthesis frequently leads to carboxylic esters (for example, see Knorr synthesis on p. 213).

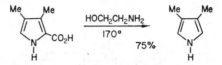

Displacement by electrophilic reagents is common. Nitration of pyrrole-2-carboxylic acid by acetic anhydride-nitric acid gives a mixture of nitro-carboxylic acids, but also 10 per cent of 2-nitropyrrole. Halogenation can effect smooth displacement of carbon dioxide (cf. furan, p. 242) and the sequence shown was used in preference to thermal decarboxylation. Of particular interest is diazo-coupling, which occurs more readily by

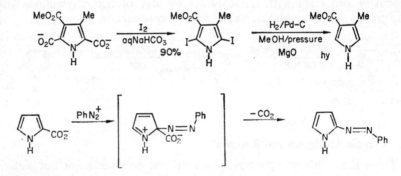

displacement of carbon dioxide than by displacement of hydrogen.

M Pyrrolecarboxylic Esters

The electrophilic substitution reactions of these stable compounds have been much studied. The main point of interest is that the normally dominant α-substitution is overcome by the directing effect of the methoxycarbonyl group in methyl pyrrole-2-carboxylate, and the electrophile enters

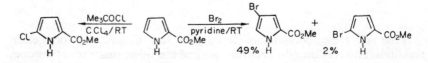

mostly in position 4—'*meta*' to the ester function. In one of the very few clearly free radical substitutions in the pyrrole field, *tert*-butyl hypochlorite gives the 5-chloro-derivative, as expected.

Also of interest is the halogenation or oxidation of alkyl groups '*ortho*' or '*para*' to the methoxycarbonyl. These reactions have found great utility in synthesis.

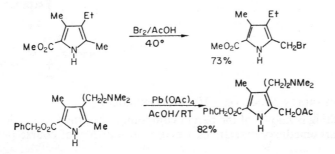

N   Halopyrroles

For reasons which are not yet clearly understood simple 2-halopyrroles are very unstable compounds. By contrast, 3-halopyrroles are quite stable (as are 2-halopyrryl ketones and esters). However, both have normal aryl halide type, relatively inert C-halogen bonds, and thus 2-chloropyrrole is unaffected by sodium, *tert*-butoxide, lithium aluminium hydride or even sodium-liquid ammonia. Catalytic hydrogenolysis occurs easily and halogen has been used as a blocking group in the pyrrole field.

O   Oxy- and Aminopyrroles

1   3- AND 4-PYRROLIN-2-ONES

If simple 2-hydroxypyrroles exist at all, it is only as extremely unfavoured tautomers of the two isomeric pyrrolones. 3-Pyrrolin-2-ones are quite

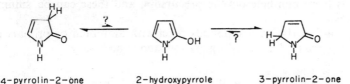

4-pyrrolin-2-one   2-hydroxypyrrole   3-pyrrolin-2-one

stable and well-known compounds; the parent compound is the major component in a tautomeric equilibrium with 4-pyrrolin-2-one and the isomers cannot be separated. Most of the known reactions of pyrrolin-2-ones are base catalysed and lead to substitution at C 5 or on oxygen, and probably proceed through a mesomeric anion as indicated.

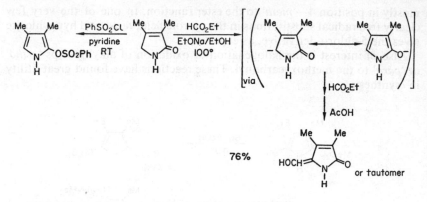

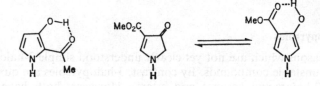

2   4-Pyrrolin-3-ones or 3-Hydroxypyrroles

Nothing is known of the parent compound or of the C-alkyl derivatives. The carbomethoxy derivatives are known, but they of course provide no

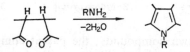

information on the possible tautomeric equilibrium of the unsubstituted system.

3   2- and 3-Aminopyrroles

Very little is known about 2-aminopyrroles beyond that they are very unstable. However, the 3-isomers appear to be much more stable, but here again much remains to be done.

P   Synthesis of Pyrrole Compounds

There are three generally important approaches to pyrrole compounds from non-heterocyclic precursors, and these can be summarized as shown.

(a) 1,4-Dicarbonyl compounds react with ammonia or primary amines to give pyrroles. No dehydrogenation step is necessary.

(b) α-Aminoketones react with carbonyl compounds which have an α-methylene grouping, preferably further activated, for example, by an ester as in the diagram.

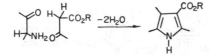

(c) Another way of putting a pair of two carbon units together employs an α-halocarbonyl compound, a keto-ester and ammonia.

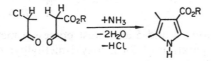

(a) *From 1,4-dicarbonyl compounds and ammonia or primary amines.*
Paal-Knorr Synthesis. Pyrroles are formed by the reaction of ammonia or a primary amine with a 1,4-dicarbonyl compound. Nucleophilic addition of the amine to the two carbonyl carbon atoms and the loss of two moles of water represents the net course of the synthesis shown

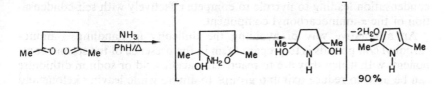

sequentially in the example. There is, however, no evidence to show in what precise order the various steps occur and no intermediates have been isolated. Note that no final oxidative step is necessary (cf. pyridine syntheses, p. 73).

A still useful synthesis of N-alkylpyrroles, which consists of dry distillation of the alkylammonium salt of mucic or saccharic acids, probably also proceeds by way of a 1,4-dicarbonyl intermediate. The overall

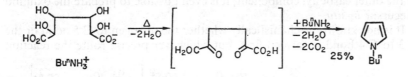

process involves the loss of four moles of water and two moles of carbon dioxide, and may proceed as shown. It is worth repeating that pyrrole carboxylic acids are easily decarboxylated.

(b) *From α-aminocarbonyl compounds.*
Knorr Synthesis. This, the most widely used general approach to pyrroles, utilizes two components: one, the α-aminocarbonyl component, supplies the nitrogen and carbon atoms 2 and 3, and the second component

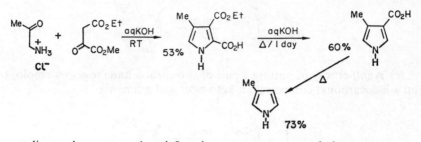

supplies carbon atoms 4 and 5 and must possess a methylene group α to carbonyl. The preparation of 2-carboxy-3-methylpyrrole, and therefrom 3-methylpyrrole, illustrates the process.

Since free α-aminocarbonyl compounds dimerize very readily to di-hydropyrazines (p. 140), they must be prepared and used in the form of their salts, only to be liberated for reaction by the base present in the reaction mixture.

The Knorr synthesis works well only if the methylene group of the second component is further activated (e.g. as in acetoacetic ester) to enable the condensation leading to pyrrole to compete effectively with self-condensation of the α-aminocarbonyl component.

An alternative way of avoiding the difficulty in handling α-amino-carbonyl compounds is to prepare them in the presence of the second component, with which they are to react. Zinc-acetic acid or sodium dithionite can be used to reduce oximino groups to amino while leaving ketone and ester groups untouched. In the following example this approach is illustrated.

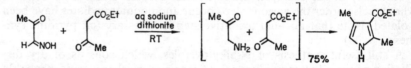

As illustrated by the classical synthesis of Knorr's pyrrole, in the special case when the α-aminocarbonyl component is simply an amino-derivative of the other carbonyl component, it is even possible to prepare the oximino precursor *in situ*.

It is not known mechanistically whether the nitrogen to C 5 bond or the C 3 to C 4 bond is formed first, but whichever precise route the reaction

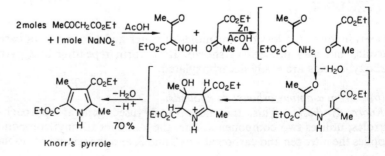

Knorr's pyrrole

takes, where there is a choice, the pyrrole formed is the result of interaction of the amino group with the most electrophilic carbonyl group of the other component. Similarly, the C 3–C 4 bond is made to the more electrophilic carbonyl group of the α-aminocarbonyl component. A reasonable sequence of events is shown for the classical synthesis.

Do not confuse this synthesis of pyrroles with the synthesis of pyridines described on p. 76.

(c)   From α-halocarbonyl compounds

Hantzsch Synthesis.   In this modification of the Feist-Benary synthesis of furans (p. 252) ammonia is incorporated into the ring system.

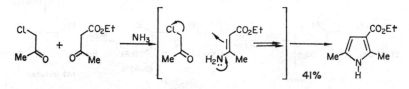

EXAMPLES OF NOTABLE RECENT SYNTHESES OF PYRROLE COMPOUNDS

(a)   *Porphobilinogen,* from 2-methoxy-4-methyl-5-nitropyridine (see p. 81). Example of Reissert indole synthesis (see p. 284).

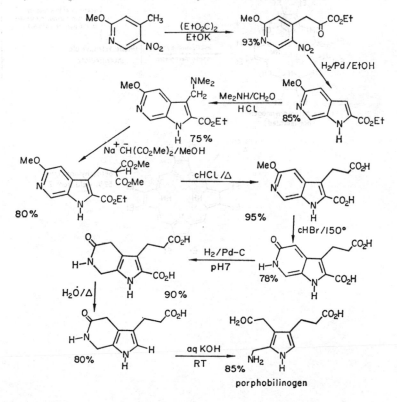

porphobilinogen

(b) *Octaethylporphyrin.* A convenient synthesis of this widely used model compound, using a Knorr synthesis as the first step, followed by appropriate modifications of substituents to give the key intermediate 2.

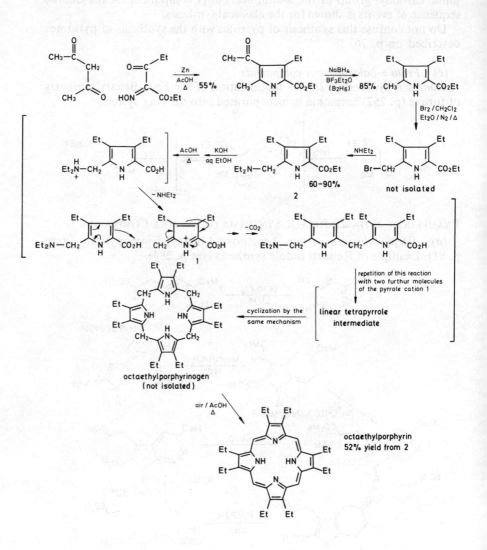

octaethylporphyrinogen
(not isolated)

octaethylporphyrin
52% yield from 2

(*c*) *Metacycloprodigiosin*, red pigment from *Streptomyces longisporus ruber*.

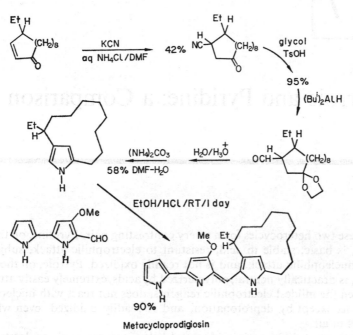

Metacycloprodigiosin

(*d*) *Methyl 4-methylpyrrole-2-carboxylate*, leaf cutting ant trail pheromone.

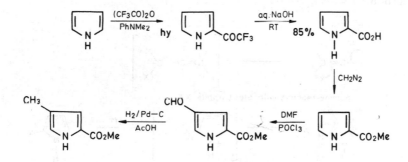

(*e*) See also p. 31.

17

Pyrrole and Pyridine: a Comparison

These two heterocycles form a very contrasting pair. Pyridine, on the one hand, is basic, stable to acid, resistant to electrophilic attack, subject to easy nucleophilic attack, and is not readily oxidized. Pyrrole, on the other hand, is practically neutral, polymerized by acids, extremely easily attacked by even the mildest electrophilic reagents, does not react with nucleophilic reagents except by deprotonation, and is readily oxidized, even when it stands in air.

These great differences in chemical behaviour are easily understood in terms of their structures, which were discussed in Chapter 1.

In most of its reactions, pyrrole behaves like an enamine: the nitrogen is in a position to release its electron pair very easily to make it available to carbon.

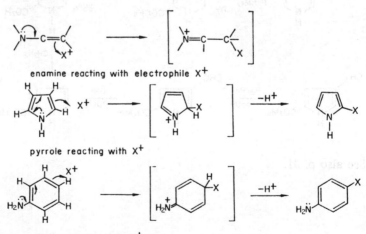

enamine reacting with electrophile X⁺

pyrrole reacting with X⁺

aniline reacting with X⁺

In this it is very closely similar to arylamines, which likewise react extremely readily with electrophiles.

Pyrrole, then, is in the same reactivity range as aniline, is much more reactive than benzene towards electrophilic reagents, and resembles both these compounds in its resistance to nucleophilic addition or substitution.

Pyridine, in complete contrast, shows a reactivity which puts it close to ketones and azomethines, functional groups which contain carbon doubly-bound to either oxygen or nitrogen.

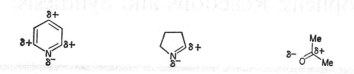

In all these compounds, polarization occurs to give a partial negative charge on the hetero atom; the polarizability also moves in the same direction, so that the carbon end of the double bond is susceptible to nucleophilic addition.

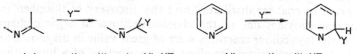

imine reacting with nucleophile Y⁻   pyridine reacting with Y⁻

Pyrroles can be likened to enamines and enols, and pyridines to the other sides of the familiar tautomeric equilibria, to azomethines and keto-carbonyls:

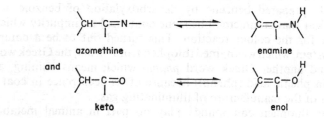

Thiophens: Reactions and Synthesis

The simple thiophens are stable liquids which closely resemble the corresponding benzene derivatives in boiling points and even in smell. They occur in coal tar distillates, and the discovery of thiophen in coal tar benzene provides one of the classic anecdotes of organic chemistry. In the early days colour reactions were of great value in diagnosis, and an important one for benzene involved the production of a blue colour on heating with isatin and concentrated sulphuric acid. In 1882, during a lecture-demonstration by Viktor Meyer before an undergraduate audience, this test failed, no doubt to the delight of everybody except the professor, and especially except the professor's lecture assistant. An inquiry revealed that the lecture assistant had run out of commercial benzene and had quickly prepared benzene by decarboxylation of benzoic acid: it was thus clear that commercial benzene contained an impurity which was responsible for the colour reaction. This turned out to be a completely new ring system, which was named thiophen from *theion*, the Greek word for sulphur, and another Greek word *phaino* which means shining, a root first used in phenic acid (phenol) because of its occurrence in coal tar, a by-product of the manufacture of illuminating gas.

Aromatic thiophen compounds play no part in animal metabolism: biotin, one of the vitamins, is a tetrahydrothiophen, and hence an aliphatic thioether (see p. 338). Thiophens, however, do occur in plants, if in a rather narrow sector, in association with polyacetylenes with which they are biogenetically closely linked (see p. 236). Banminth (Pyrantel), a valuable anthelminth used in animal husbandry, is one of the few thiophen compounds in chemotherapy.

Thiophen is manufactured by the gas-phase interaction of C 4 hydrocarbons and elementary sulphur at 600°.

REACTIONS AND SYNTHESIS OF THIOPHENS

A   Reactions with Electrophilic Reagents

1   SUBSTITUTION AT CARBON

(a) *Protonation.* Thiophen is stable to all but very vigorous acidic conditions, so that many reactions which lead to acid-catalysed decomposition or polymerization with furans and pyrroles can be carried out on thiophens, which thus resemble benzene more closely.

Deuteration experiments in aqueous sulphuric acid show that protonation of thiophen at the α-positions occurs about 1000 times faster than that of benzene under the same conditions. The thiophen β-positions are much less reactive and are protonated at about the same rate as benzene.

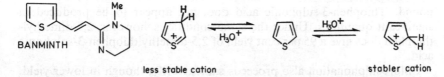

BANMINTH

less stable cation   stabler cation

Alkyl substitution greatly facilitates protonation, so that whereas thiophen is not deuterated at all in boiling deuteroacetic acid, 3-methylthiophen gives 2-deutero-3-methylthiophen after a few hours. That the other α-position, C 5, is not deuterated is noteworthy, and parallels the behaviour of 3-methylfuran (see p. 240).

The action of 100 per cent phosphoric acid on thiophen leads to a trimer. Its structure suggests that, in contrast with pyrrole, the electrophile involved in the initial step is the α-protonated cation.

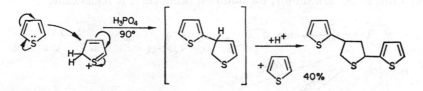

40%

(b) *Nitration.* Thiophen reacts violently, sometimes explosively after an induction period, with normal nitrating mixtures, even the relatively mild $AcOH-HNO_3$; this is considered to be due to an autocatalytic nitrosation. Much milder reagents give a 6:1 mixture of 2- and 3-nitrothiophens. The highest yield of mononitro product, 91 per cent, has been reported by using nitronium borofluoride in ether.

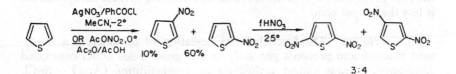

3:4

2-nitrothiophen is much less reactive and can be smoothly nitrated with fuming nitric acid.

(c) *Sulphonation.* The smooth and rapid sulphonation of thiophen at room temperature provides the basis for its isolation from coal tar benzene. The thiophen-2-sulphonic acid thus formed is easily separated and hydrolysed back to thiophen by superheated steam. The highest yield of thiophen-2-sulphonic acid is obtained with the pyridine sulphur trioxide com-

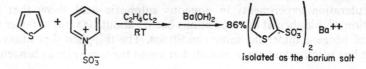

isolated as the barium salt

pound. Thiophen-3-sulphonic acid does not appear to be produced in significant quantities. Under the same conditions, however, 2,5-dimethylthiophen does give a 95 per cent yield of 2,5-dimethylthiophen-3-sulphonic acid.

Chlorosulphonation also proceeds satisfactorily, though in lower yield.

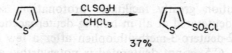

37%

(d) *Halogenation.* This occurs very readily at room temperature, and is rapid even at $-30°$ in the dark. In acetic acid at room temperature thiophen reacts with chlorine 10^7 times faster and with bromine 10^9 times faster than does benzene. The main products of the action of one mole of chlorine at 50° are shown; the addition product, 1, is remarkable.

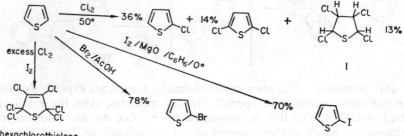

hexachlorothiolene

With excess chlorine and a catalytic quantity of iodine a quantitative yield of the hexachlorothiolene is obtained. In great contrast with nitration, the proportion of 3-monosubstitution occurring in these halogenations is less than 1 per cent.

(e) *Acylation.* The Friedel-Crafts acylation of thiophen is a much-used reaction and proceeds generally to give good yields under controlled conditions; it leads almost exclusively to α-substitution (Ac₂O — SnCl₄ in $C_2H_4Cl_2$ at RT → α:β ratio of 200:1). Anhydrous aluminium chloride and, to a lesser extent, stannic chloride react with thiophen to produce intractable tars. However, this undesirable resinification is largely avoided by adding the catalyst gradually to a mixture of the thiophen and the

acylating agent, when the aluminium chloride reacts preferentially with the acylating agent to give the active electrophile.

The resinification of thiophen is the reason why thiophen-free benzene must be used in Friedel-Crafts reactions.

Acylation with carboxylic anhydrides may be catalysed by strong acids, thus a good route to 2-acetylthiophen involves the use of acetic anhydride and phosphoric acid.

Vilsmeier formylation proceeds readily.

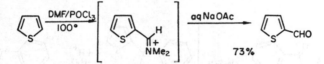

(*f*) *Alkylation* occurs readily, but generally it is not a useful procedure in preparation. The highly reactive, and consequently less selective, carbon-

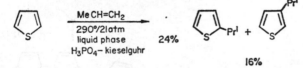

ium ions discriminate only poorly between α- and β-positions; poly-alkylation and resinification are difficult to avoid. (See p. 226 for S-alkylation.)

(*g*) *Condensation with aldehydes and ketones.* Acid-catalysed reaction of thiophen with aldehydes and ketones is not a good route to hydroxy-

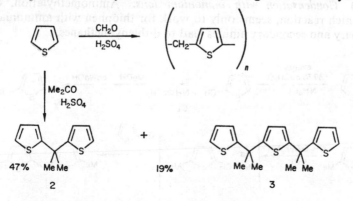

alkylthiophens, for these are unstable under the reaction conditions. Thiophen and formaldehyde give polymeric products, but the reaction with acetone can be controlled to give 2 and 3.

Chloromethylation is a valuable reaction in synthesis but seems to be very dependent on precise reaction conditions.

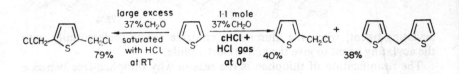

A reaction of special historical interest, mentioned in the introduction, is the condensation of thiophen with isatin in concentrated sulphuric acid to give the deep blue indophenine.

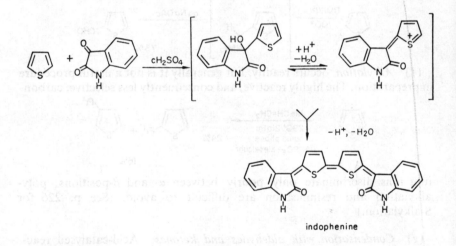

indophenine

(h) *Condensation with immonium ions.* Aminomethylation, or the Mannich reaction, seems only to work for thiophen with ammonia itself. Primary and secondary amines lead to dithienylmethanes.

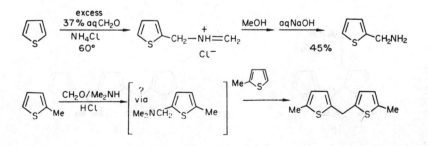

(*i*)   *Mercuration* occurs with great ease. Mercuric acetate is appreciably more reactive than mercuric chloride, as the examples demonstrate.

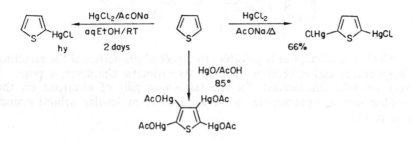

Even powerfully deactivating groups hardly diminish the ease with which this substitution occurs.

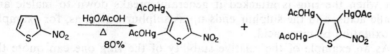

(*j*)   *Diazo-coupling.* Thiophen itself is insufficiently reactive to couple, even with reactive diazonium salts such as the 2,4-dinitro diazonium ion, only 2-mono- and 2,5-diarylation is observed. 2-*tert*-Butylthiophen on the other hand does couple normally with this diazonium salt. 2,5-Dimethyl-thiophen gives the products of normal coupling on the ring and abnormal coupling at a side chain methyl group.

Me₃C $\diagdown$ (NO₂)₂C₆H₃N₂⁺/RT/1 day   Major product Me₃C—S—N=N—C₆H₃(NO₂)₂
        AcOH—H₂SO₄

2   DISPLACEMENT OF GROUPS AT α-POSITIONS

Thiophen compounds provide many examples of this type of substitutive displacement; the following two examples are typical.

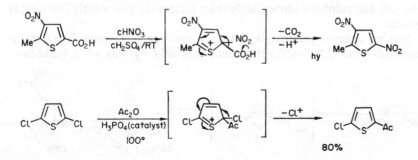

3 Addition to Ring Sulphur

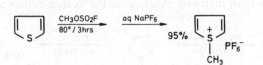

Alkylation of sulphur is possible: the spectral properties of the resulting thiophenium cation show it to retain its aromatic character, a point of very considerable interest, for the remaining pair of electrons on the sulphur can still participate in the aromatic π-molecular orbital system (see p. 11).

B Reactions with Oxidizing Agents

The thiophen ring system can survive moderate oxidizing conditions, but when the ring is attacked it generally breaks down to maleic and oxalic acids, and the sulphur ends up as sulphuric acid, as, for example, in oxidation by nitric acid.

As an example of the relative stability of the ring one can quote the extraordinary base-catalysed autoxidation of alkyl groups to give thiophen carboxylic acids.

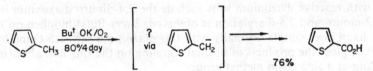

Percarboxylic acids effect specific oxidation of the sulphur, the highly reactive, non-aromatic sulphones only being isolable in cases where the ring is substituted. In the case of thiophen itself, the intermediate sul-

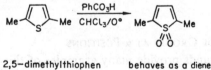

2,5–dimethylthiophen   behaves as a diene with maleic anhydride

phoxide and sulphone immediately react further, to give mostly intractable products.

Singlet oxygen (photoactivated via sensitizer methylene blue) reacts smoothly with 2,5-dimethylthiophen, probably by way of 1,4-addition.

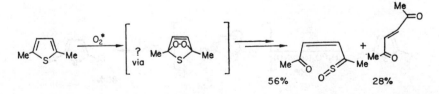

C   Reactions with Nucleophilic Reagents

Thiophen and alkylthiophens do not react with nucleophilic reagents by substitution or addition. As with furans and N-alkylpyrroles, they only react with the strongest bases, and then by deprotonation at an α-carbon.

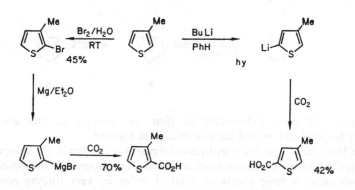

Butyllithium reacts in this way to give synthetically useful thienyllithium reagents. Because the factors determining the position of electrophilic substitution in substituted thiophens are quite different from those determining the site of deprotonation, these two reaction types can at times be used to obtain isomeric products, as the scheme illustrates.

In competing reaction, thiophen is found to be C 2 deprotonated by *tert*-butyllithium 25 times more rapidly than furan.

D   Reactions with Free Radicals

Thiophen reacts with phenyl radicals about three times as fast as does benzene and very largely at an α-position.

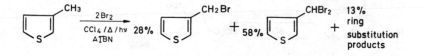

+ some 3-isomer

Good yields of 2-acyloxy-derivatives can be similarly obtained.

Side-chain, free-radical bromination can be made to compete, moderately successfully, with ring electrophilic halogenation by adding bromine slowly to the alkylthiophen in the presence of a radical initiator and light.

E Reactions with Reducing Agents

The thiophen ring is not normally reduced by metal-acid combinations or by hydride reagents. Although unattacked by sodium at 80°, thiophen is reduced by sodium-liquid ammonia-ethanol to give two dihydrothiophen products, 4 and 5. Sodium in methanol reduces more

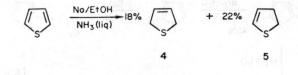

4   **5**

slowly but more extensively so that in addition to the above two dihydroproducts, mixed butene thiols are formed.

Clean reduction to tetrahydrothiophens is difficult to achieve catalytically because of the tendency of sulphur compounds to poison catalysts. With either a large excess of catalyst or under very forcing conditions, tetrahydrothiophen can be obtained in good yield. Except for thiophen itself, where poor yields are obtained, a recently devised method involving the triethylsilane reduction of a protonated thiophen is a good method for the formation of tetrahydrothiophens.

Raney nickel is of special interest for it achieves reductive desulphurization of thiophens in high yields, and thus provides a very useful synthetic route to many aliphatic compounds, as the following example shows.

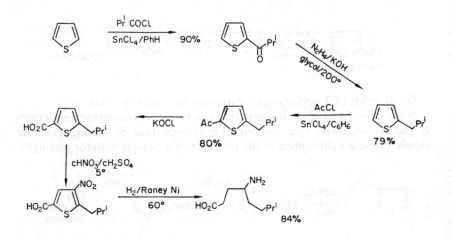

The desulphurization may also be effected very simply by dissolving Raney alloy in an alkaline aqueous solution of a thiophen (generally used for thiophens containing a carboxyl group).

F Reactions with Dienophiles

Though benzyne does not react with thiophen, the much more reactive tetrafluorobenzyne does react, by 1,4-addition; the only other known cycloaddition is with the very reactive dicyanoacetylene.

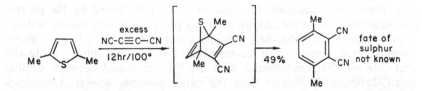

G Reactions with Carbenes

Carbethoxycarbene adds to the C 2—C 3 double bond of thiophen to give a cyclopropane compound, 6, which can be opened to give a thiophen-β-acetic ester derivative (compare p. 207, 245).

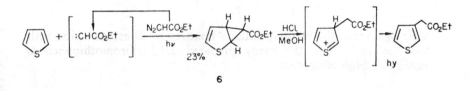

6

H Halothiophens

Halothiophens play an important role in the synthesis of thiophen compounds as intermediates for the preparation of thienyllithium and thienyl Grignard reagents.

2-Halothiophens react very readily with magnesium to give good yields of thienyl Grignards, and with lithium or lithium alkyls to give the corresponding thienyllithium compounds.

3-Iodothiophens also react with magnesium, but 3-bromothiophen only reacts by the entrainment method. On the other hand, 3-bromothiophen reacts easily and smoothly with butyllithium at −70° to give 3-thienyllithium. This reaction probably affords the best route to 3-monosubstituted thiophens, since 3-bromothiophen is readily synthesized (see below).

If the reaction between butyllithium and 3-bromothiophen is conducted at room temperature, then a complicated series of reversible reactions occur leading to thiophen, 2-thienyllithium and 3-bromo-2-thienyllithium.

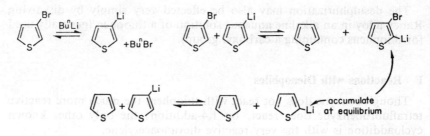

This surprising result is rationalized in the scheme shown. In these reversible-interchange reactions, the equilibrium is determined by the greater stabilization of negative charge at the α-position, which makes α-lithio compounds stabler than their β-isomers and also makes α-halothiophens more reactive in this sort of reaction than the β-halothiophens. Note also that 3-bromo-2-thienyllithium, with the lithium joined between sulphur and C-bromine, is stabler than the other possible isomer, 3-bromo-5-thienyllithium.

These interchange reactions can be visualized as involving a four-centre transition state, 7.

7

α-Halogens may also be preferentially removed by direct reduction with zinc-acetic acid, thus the easily prepared 2,3,5-tribromothiophen is reduced to 3-bromothiophen.

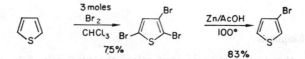

Although nothing much is known yet about the relative ease of nucleophilic substitution of simple halothiophens versus simple halobenzenes, halothiophens that contain a nitro group have been shown to react much faster with nucleophilic reagents than the corresponding benzene compounds.

Relative rates of reaction with piperidine at 25°.

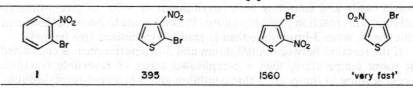

| 1 | 395 | 1560 | 'very fast' |

This greater ease of reaction is not shared with nitrohalofurans or -pyr-roles, and can be ascribed to participation of the sulphur in the delocaliza-tion of the negative charge as in the intermediate, 8, by the expansion of its shell to 10 electrons.

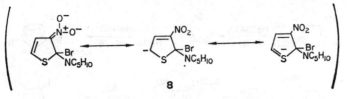

8

The cuprous-salt catalysed nucleophilic displacements are of consider-able importance in preparation, as indeed they are in most other groups of aromatic halides.

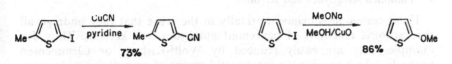

I Reactions of Metallothiophens

The most important of this class of compounds are the thienyllithium compounds and the thienyl Grignard reagents, the preparation of which has been discussed earlier. These organometallic reagents behave normally with the usual range of electrophiles, if sufficient care is taken not to allow interchange or ring-opening reactions to occur and complicate matters.

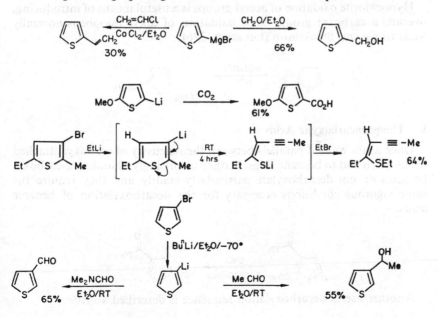

The thienylmercurihalides and acetates have also played an important role in isolation and synthesis, largely because of the ease with which such groups are introduced and then replaced either by hydrogen (i.e., removed) or by halogen.

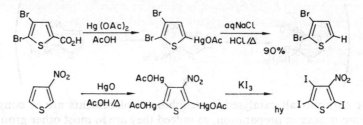

J Thiophen Aldehydes and Ketones

These compounds behave normally in the sense that they undergo all the carbonyl reactions of benzenoid aldehydes and ketones. Thus, for example, they are easily reduced by Wolff-Kishner or Clemmensen methods, which provide the most useful means of preparing a wide range of alkyl thiophens. The Friedel-Crafts alkylation is not sufficiently selective to be widely useful in synthesis.

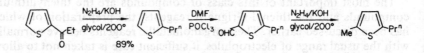

Hypochlorite oxidation of acetyl groups is a useful means of introducing, overall, a carboxyl group, since oxidation of alkyl thiophens normally leads to general breakdown (but see p. 226).

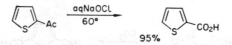

K Thiophencarboxylic Acids

Once again we find parallels between the reactivity of carboxyl attached to thiophen and to benzene rings. Rather surprisingly, thiophen-2-carboxylic acids do not decarboxylate particularly readily and they require the same vigorous conditions necessary for the decarboxylation of benzoic acids.

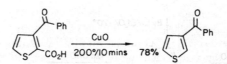

Another useful decarboxylation sequence is described above.

L   Oxy- and Amino-Thiophens

1 Hydroxythiophens or Thiolenones

These compounds are much more difficult to handle and much less accessible than the phenols. The '2-hydroxythiophens' are stable in the absence of oxygen and exist mainly in one or both of the carbonyl tautomeric forms, and are better described as thiolenones.

Thiol-3-en-2-one is the main tautomer, neither of the other two, 2-hydroxythiophen and thiol-4-en-2-one, is detectable spectroscopically.

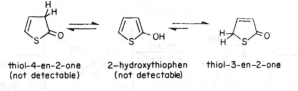

thiol-4-en-2-one   2-hydroxythiophen   thiol-3-en-2-one
(not detectable)   (not detectable)

A methyl group at C 5 stabilizes a C 4—C 5 double bond so that in this compound 5-methylthiol-3-en-2-one is in equilibrium with about 15 per cent of the tautomeric 5-methylthiol-4-en-2-one.

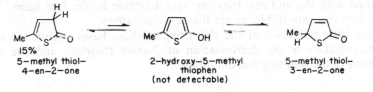

5-methyl thiol-   2-hydroxy-5-methyl   5-methyl thiol-
4-en-2-one   thiophen   3-en-2-one
15%   (not detectable)

By contrast the much less stable 3-hydroxythiophen appears to exist largely as the hydroxy tautomer.

2-hydroxythiophen   thiol-4-en-3-one

The thiolenones are prepared by the action of oxygen on thienyl Grignard reagents or hydrogen peroxide on thienylboronic acids, and are not prepared by methods normally used for the synthesis of phenols.

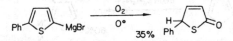

Surprisingly, nothing seems to be known about the acidity of these compounds.

Electrophilic reagents react either at oxygen or at C 5, depending on the reagent. O-acetylation and O-methylation may well proceed *via* the anion, since they are carried out in aqueous alkaline solution, whereas the diazocoupling and the condensation with benzaldehyde probably proceed *via* a low equilibrium concentration of hydroxy tautomer.

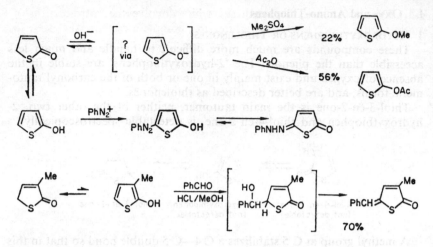

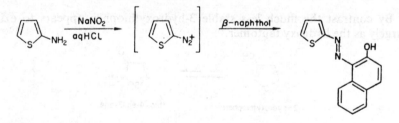

2 AMINOTHIOPHENS

The 2- and 3-aminothiophens exist as the amino tautomers but, in contrast with the anilines, they are very unstable as the free bases. The salts, however, are stable, as are the acyl-derivatives.

Very little is known of the chemistry of these bases; about the only simple reaction is the diazotization of 2-aminothiophen and the azo-coupling of the resulting diazonium ion.

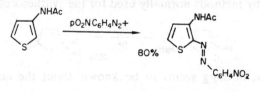

The thiophen ring is so strongly activated by an amino group that even acetamidothiophen couples with *p*-nitrobenzene diazonium ion.

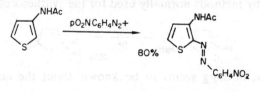

M Synthesis of Thiophen Compounds

1 RING SYNTHESIS

The general ways of synthesizing thiophen rings from non-heterocyclic precursors can be broadly divided into two categories.

(a) First there are the syntheses which involve the introduction of sulphur into a precursor containing the complete carbon skeleton. In

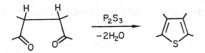

general this means having a 1, 4-dicarbonyl component or an oxidation level equivalent, or reaction conditions which lead *in situ* from a reactant at a lower oxidation level to a suitably oxidized material.

(b) Secondly there are those which employ either the reaction of a

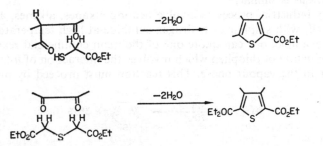

mercapto acetate with a 1,3-dicarbonyl compound or the reaction of a thiodiacetate with a 1,2-dicarbonyl compound.

(a) *From 1,4-dicarbonyl compounds.* The reaction of a 1,4-dicarbonyl compound with phosphoric or phosphorous sulphides to give thiophens was, until more recent times, the main route in the laboratory. Where this

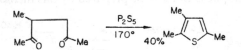

cyclization differs radically from the corresponding furan and pyrrole ring syntheses is in the ability of phosphorus sulphides to reduce a carboxyl

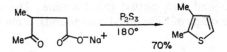

function (if part of the dicarbonyl component) eventually to give an α-CH. Precisely at what stage this reduction occurs is not yet known.

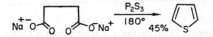

One of the recently-developed modifications of this approach makes use of diacetylenes. These are, of course, at the same oxidation level as 1,4-diketones, and have been found to react with hydrosulphide ion under mild conditions. Nearly all the naturally-occurring thiophens are found in

$$Et.C{\equiv}C-C{\equiv}C.Et \xrightarrow[\text{I day /RT}]{\text{NaSH/EtOH}} \text{65\%} \quad \text{Et}\underset{S}{\diagup\!\!\!\diagdown}\text{Et}$$

plant genera which normally produce polyacetylenes, and in view of the above easy *in vitro* reaction, it seems reasonable to suppose that their biosynthesis is similar.

Many industrial processes involve heating alkanes, alkenes, and acetylene itself, with sulphur or hydrogen sulphide at high temperatures. As a simple example, one can quote one of the main commercial reactions for the production of thiophen which involves the interaction of *n*-butane and sulphur in the vapour phase. This reaction must proceed by prior dehy-

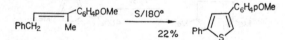

20% per pass; 70% by recycling

drogenation of the butane, followed by addition of sulphur to the unsaturated intermediates. It is, of course, well known that butadiene can be produced by sulphur dehydrogenation of butane.

This type of thiophen synthesis only works with low molecular weight hydrocarbons, for cracking processes interfere when one goes beyond hexanes. Aryl thiophens can also be prepared by this method.

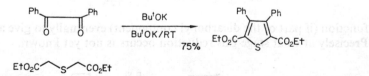

(*b*) *From thiodiacetates and mercapto acetates.*

(i) *From Thiodiacetates and 1,2-Dicarbonyl Compounds. The Hinsberg Synthesis* involves two consecutive aldol condensations between a 1,2-dicarbonyl compound and diethyl thiodiacetate. The approach can be applied to a wide range of dicarbonyl compounds, including oxalic esters.

(ii) *From Mercapto Acetates and 1,3-Dicarbonyl Compounds.* Mercapto

acetic esters may also be used with a very wide range of bifunctional compounds such as 1,3-dicarbonyl compounds and derivatives, or conjugated
acetylenic esters or ketones. The reaction proceeds by nucleophilic addition
of the thiolate anion, followed by a cyclic Claisen condensation.

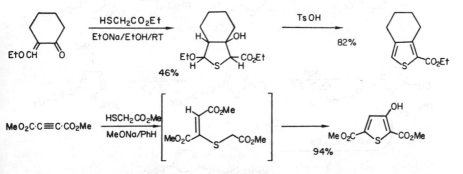

2 SYNTHESES STARTING FROM THIOPHEN OR SIMPLE ALKYL THIOPHENS

The introduction and elaboration of substituents in simple commercially-
available thiophens is probably resorted to more often than the corresponding conversions in pyrrole and furan synthesis. This is largely because of
the greater resistance of the thiophen ring to undesired side reactions under
most conditions. In this respect thiophen chemistry stands close to benzene
chemistry, for example, as in benzene chemistry, the introduction of a
directing group followed by its removal can be used to introduce a substituent into an otherwise unfavoured position.

(a) Tetrahydrobenzo[b]thiophen

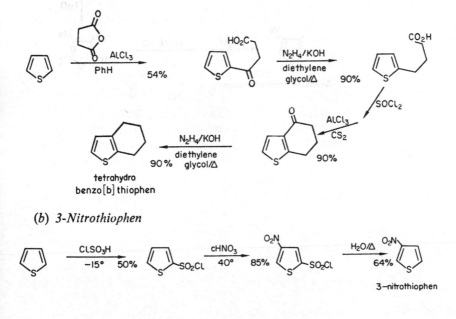

(b) 3-Nitrothiophen

(c) *Junipal*, from a wood-rotting fungus, *Daedalea juniperina* and its conversion into the ester, 9, from *Chrysanthemum vulgare*.

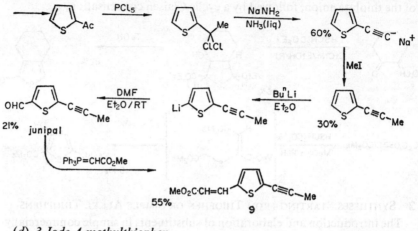

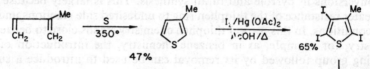

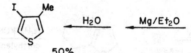

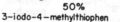

55%

9

(d) *3-Iodo-4-methylthiophen.*

50%
3−iodo−4−methylthiophen

Furans: Reactions and Synthesis

Furans are volatile, fairly stable compounds with pleasant odours. Furan itself is slightly soluble in water. It is readily available, and its commercial importance is mainly due to its role as the precursor of the very widely used solvent tetrahydrofuran. Furan is produced by the gas-phase decarbonylation of furfural (2-formylfuran), which in turn is prepared in very large quantities by the action of acids on vegetable residues mainly from the manufacture of porage oats and cornflakes. Furfural was first prepared in this way as far back as 1831 and its name is derived from *furfur*, which is the latin word for bran: in due course, in 1870, the word furan was coined from the same root.

The aromatic furan ring system, though not found in animal metabolism, occurs widely in secondary plant metabolites, especially in terpenoids: perillene is a simple example. The occurrence of furfurylthiol in the aroma

perillene

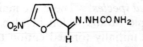

furfuryl thiol

of roasted coffee is of some interest, for it is not present in the fresh bean. A number of 5-nitrofurfural derivatives are important chemotherapeutic agents, nitrofurazone, a bactericide, being a simple example.

nitrofurazone

REACTIONS AND SYNTHESIS OF FURANS

A   Reactions with Electrophilic Reagents

(a) *Protonation.*   Furan and the simple alkyl furans are stabler towards mineral acids than is generally believed. Although furan is instantly decomposed by concentrated sulphuric acid or by strong Lewis acids such as aluminium chloride, it reacts only slowly with HCl either as the ordinary concentrated acid or as a solution in a non-hydroxylic solvent. Hot dilute aqueous mineral acids, however, cause hydrolytic ring-opening (see below).

Although the pK_a of furan has not been measured, it is much lower than that of aliphatic ethers, as is evident from the fact that diethyl ether dissolves extensively in concentrated aqueous hydrochloric acid, whereas furan does not. Acid-catalysed deuteration occurs at the α-position by way of cation 1; an analogous deuteration at β-positions, which would involve cation 2, has not been observed. A much slower formation of cation 2 is however believed to occur, but this cation is so much more reactive than cation 1, that it adds water, leading to hydrolytic ring-opening (see below). Whether or not cation 3 is also present in the protonation equilibrium is not known.

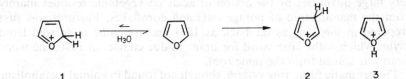

1   2   3

Alkyl groups can exert a powerful directing and stabilizing effect as will be seen in some of the substitution reactions which follow, and in the specific deuteration of 3-methylfuran at C 2, which shows protonation at the other α-position, C 5, to be a very much slower process. More effective

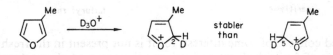

conjugation of the methyl with the mesomeric cation must be the main factor controlling this specific deuteration.

Reactions of protonated species.   These are mainly seen in the hydrolysis and alcoholysis of simple furans which involve nucleophilic addition of water or alcohol to the initially formed cation to give derivatives of 1,4-dicarbonyl compounds. This is, in effect, a reversal of one of the general syntheses of furans.

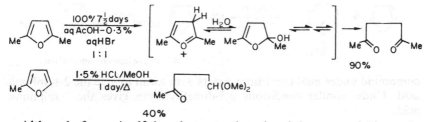

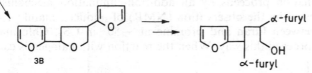

40%

Although furan itself has been methanolysed in poor yield to the succindiacetal, it has not yielded succindialdehyde on hydrolysis, presumably because of the instability of the latter under the reaction conditions.

The resinification of furans in mineral acids may well involve polymerization by way of C-protonated species, as in pyrroles.

Furan is converted in low yield into two tetrameric products by concentrated hydrochloric acid under mild conditions for several hours. One product arises from an intermediate trimer 3A derived from cation 2 (cf. pyrrole trimer, p. 201) and the other from the isomeric trimer 3B, derived from cation 1.

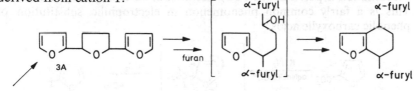

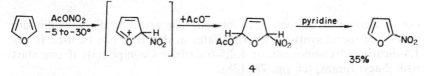

(b) *Nitration* is achieved with the mild nitrating agent acetyl nitrate. The special feature of this nitration is that a more-or-less stable 2,5-adduct, 4, is formed, which eliminates acetic acid either thermally or by the agency of pyridine to give 2-nitrofuran.

2-Nitrofuran has also been obtained by the action of nitronium borofluoride in ether. Since this system contains no effective nucleophile, an intermediate 2,5-adduct is almost certainly not involved, and the intermediate cation deprotonates directly.

(c) *Sulphonation.* Furan and its simple alkyl derivatives are decomposed by the usual strong reagents, but the pyridine sulphur trioxide

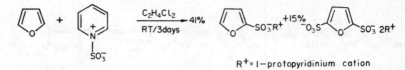

R$^+$= I—protopyridinium cation

compound under mild conditions reacts with furan to give the 2-sulphonic acid. Under similar conditions 2,5-dimethylfuran gives the 3-sulphonic acid.

(d) Halogenation. Furan reacts vigorously with chlorine and bromine at room temperature to give polyhalogenated products, but does not react at all with iodine. Much milder conditions have to be used to obtain monochloro- or monobromo-derivatives.

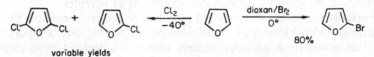

variable yields

Iodination is possible when displacement of carbon dioxide is involved. This is a fairly common phenomenon in electrophilic substitution of phenolic carboxylic acids.

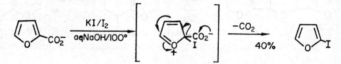

That bromination proceeds by an addition-elimination mechanism is strongly suggested by the observation (NMR) of adducts, mainly 1,4, in the reaction between furan and bromine at $-50°$ in CS_2. Chlorination may well also proceed this way. When the reaction with halogen is carried

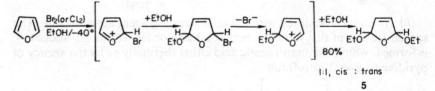

1:1, cis : trans

5

out in alcohol an adduct, 5, is formed. The resulting dialkoxy-dihydro-furans are useful synthetic intermediates in that they can be used to give 1,4-dicarbonyl compounds or 1,2,5-tricarbonyl compounds if one starts with 2-acyl furans. (cf. pp. 78, 138).

(e) Acylation. Carboxylic anhydrides or carbonyl halides normally react only in the presence of Friedel-Crafts or of orthophosphoric acid catalysts, however, more reactive anhydrides, such as $(CF_3CO)_2O$ (see p. 186) can effect uncatalysed substitution. Furan itself can be acetylated with $Ac_2O–SnCl_4$ or $AcOSO_2C_6H_4pCH_3$ at RT to give 2-acetylfuran almost exclusively (the ratio of 2- to 3-substitution was shown to be

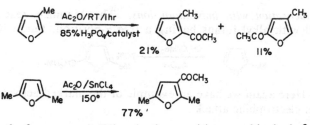

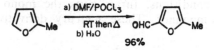

6800:1 in the former case). When both α-positions are blocked, β-acylation occurs smoothly. There is little polymerization since the Lewis acid is largely complexed with the acylating agent and then, after the reaction, with the carbonyl group of the product.

Although Gatterman formylation works well, better yields are obtained by the Vilsmeier method.

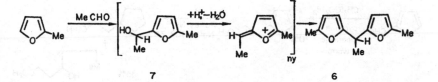

(f) *Alkylation.* The Friedel-Crafts alkylation reaction is not useful with furans and simple derivatives, partly because of polymerization by the catalyst and partly because of polyalkylation. A few isolated cases of useful alkylation are known, however; the example shown provides one of the very few instances of actual isolation of a β-monosubstituted product. As very strong electrophiles, carbonium ions are among the least selective.

Allylation occurs smoothly under the mild conditions required for the generation of allylcarbonium ions.

(g) *Condensation with aldehydes and ketones.* This occurs by acid catalysis. Furan reacts with acetaldehyde to give a separable mixture of linear oligomers, whereas 2-methylfuran gives 6 as main product. In these reactions, the intermediate alcohols, such as 7, are too reactive to be isolable under acidic conditions. Furan reacts with acetone, again to give mainly mixed linear oligomers, but also to yield 20 per cent of the macrocyclic tetramer (see pyrrole analogue, structure 1 on p. 199); the yield of the tetramer is doubled in the presence of $LiClO_4$, clearly the consequence of coordination of the furan oxygens with Li^+.

(*h*) *Condensation with immonium ions.* The Mannich reaction proceeds well with alkyl furans having a free α-position. Furan itself does

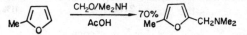

not react. Here again we have an example of the marked activating effect of CH_3 on electrophilic attack.

(*i*) *Diazocoupling and nitrosation.* Furan and mono-alkylfurans do not react with nitrous acid or with diazonium salts, even the more reactive ones. However, 2,5-dimethylfuran does react with *p*-nitrobenzene diazonium chloride, though in a complex manner.

(*j*) *Mercuration.* Reaction with mercuric salts occurs very readily. Depending upon conditions, furan gives the mono- through to the tetrasubstituted compounds. These mercury derivatives are useful in synthesis.

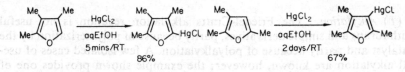

B Reactions with Nucleophilic Reagents

Furan and alkyl furans do not react with nucleophiles by addition or by substitution. The stronger bases, however, can effect deprotonation at the

α-position. Thus, lithium *t*-butoxide in dimethyl sulphoxide/O-deutero-*t*-butanol brings about deuteration of both α-positions. Under non-equilibrating conditions, α-metallation occurs with lithium alkyls or with sodium aluminium hydride, but not with lithium aluminium hydride, Grignard reagents, or sodium or potassium metals.

C Reactions with Reducing Agents

The best way to reduce a furan to a tetrahydrofuran is by catalysis with Raney nickel, even if sometimes hydrogenolysis products are also produced.

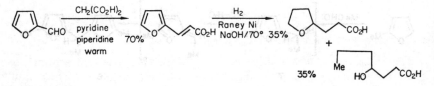

Metal halides, dissolving-metal combinations, or sodium in liquid ammonia do not reduce furan or simple alkyl furans. The much more powerful lithium/methylamine/methanol does reduce them, but the reaction is complex as the example shows.

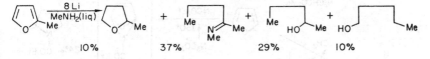

In contrast with alkyl furans, furan-2-carboxylic acid is reduced by the sodium-liquid ammonia-methanol combination to the 2,5-dihydro derivative.

D   Reactions with Radical Reagents

Furan reacts with free radicals at the α-position. Phenyl radicals generated by a variety of means, for example decomposition of benzene diazonium ion, react with furan to give 2-phenylfuran.

Decomposition of dibenzoyl peroxide in the presence of furan gives a good yield of the *cis* and *trans* isomers of the addition product 8.

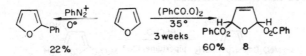

E   Reactions with Dienophiles

Furan reacts as a diene with powerful dienophiles like maleic anhydride and benzyne, to give Diels Alder adducts. No reaction occurs, however,

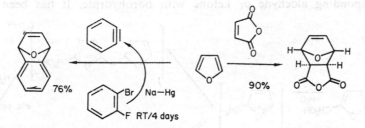

with slightly less reactive dienophiles like acrolein or methyl acrylate unless the furan in question is further activated, e.g. by the presence of a methoxyl group.

F   Reactions with Carbenes and Nitrenes

The reaction of furan with carbethoxynitrene gives rise to a product, 9, probably via a 1,4-adduct as illustrated, on the other hand, carbene and

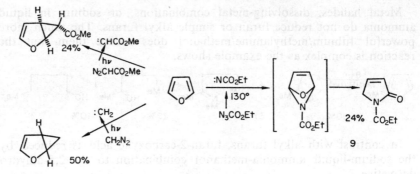

methoxycarbonyl carbene generated photolytically add to the 2,3-double bond.

G Alkylfurans

Alkyl groups on the furan ring have no special reactivity beyond that which is associated with their being benzylic. A powerfully activating ring substituent is necessary to labilize the hydrogens of such a group, just as is required in benzene chemistry.

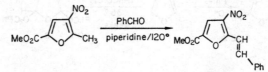

H Furan-C-X Compounds

In general furfuryl alcohols can be prepared by reduction of the corresponding aldehyde or ketone with borohydride. It has been noted

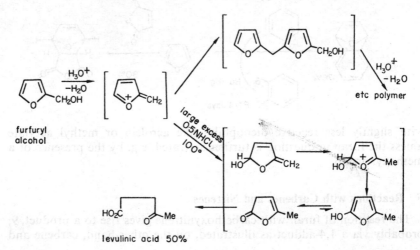

earlier that they cannot be isolated from reactions between an aldehyde or ketone and a furan. This is because they are extremely reactive in the presence of acid: rapid loss of water is promoted by acid and the resulting mesomeric ion reacts as an electrophile, and hence polymers are generally formed. In the reaction of furfuryl alcohol with an excess of dilute aqueous acid it is possible to trap the initial cation with the large quantity of nucleophile (water) present, and in this way moderate yields of levulinic acid can be obtained.

Furfuryl chloride is of course also extremely reactive, as its reaction with

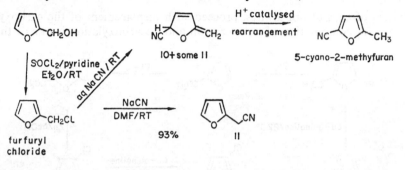

sodium cyanide shows. In aqueous solution this reaction takes the form of an allylic displacement and the isomeric 5-cyano-2-methylfuran is produced. The non-aromatic intermediate, 10, has been observed by NMR and a solution containing it can be kept without change for weeks at 0°. It aromatizes when heated to RT.

I   Furan Aldehydes and Ketones

Furan-2-aldehyde, furfural, is commercially available (see under synthesis) and, since it can readily be reduced to the alcohol or oxidized to the acid, is a key starting material for many furan syntheses.

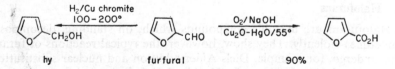

Electrophilic substitution of 2- or 3-acylfurans generally occurs at C 5. That the nitration of 2-acetylfuran gives two 1,4-addition products shows that not all of the initial attack takes place at C 5.

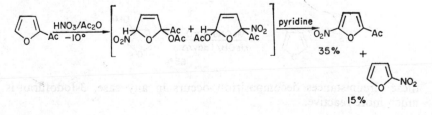

J Furoic Acids

Electrophilic substitution of furoic acids proceeds well: 3-furoic acids, like the 3-acyl furans, are substituted at C 5; the reactions of 2-carboxy-

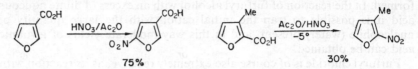

furans often, but not always, proceed with displacement of the carboxyl group. A combination of halogenation and decarboxylation is one of the

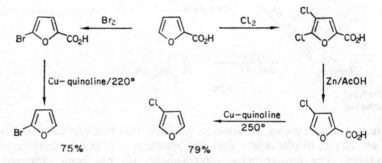

best ways to prepare halo-furans, for decarboxylation of furoic acids occurs smoothly, an α-carboxyl being lost more easily than a β-carboxyl as carbon dioxide.

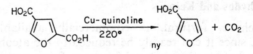

K Halofurans

Halofurans are unstable compounds which, on standing, decompose, sometimes violently. They show, however, the typical reactions of furans and undergo, for example, Diels Alder addition and nuclear substitution with acid chlorides or mercuric chloride. 2-Halofurans are not particularly susceptible to nucleophilic attack (displacement of halogen). For example, the vigour necessary to induce reaction with methoxide is comparable to that necessary to make chlorobenzene react with this reagent, and under

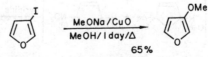

these circumstances decomposition occurs in any case. 3-Iodofuran is much more reactive.

As in benzene chemistry, the introduction of electron-withdrawing groups makes displacement much easier.

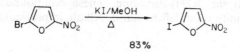

83%

The halofurans do not give Grignard or lithium compounds by reaction with magnesium or lithium under the usual conditions. It is worth repeating that the 2-lithiofurans can be made by the direct metallation of furans. 3-Metallated furans can be obtained from 3-halofurans and butyllithium. Once formed, both isomeric lithio-types behave normally and are thus extremely useful in synthesis.

L   Oxy- and Aminofurans

1   2-OXYFURANS

If 2-hydroxyfuran exists at all, it is only at an undetectably low concentration in a tautomeric equilibrium involving a 2(5H)-furanone and a 2(3H)-furanone. The angelica lactones can be equilibrated with triethyla-

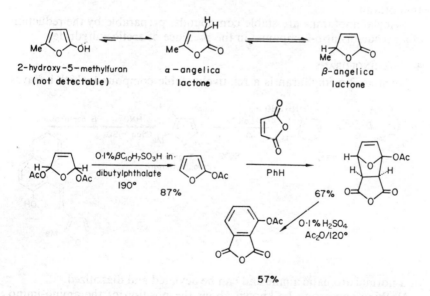

mine, the stabler being the β-isomer. The chemistry of the 2-oxyfurans, then, is essentially that of unsaturated lactones and has no special relevance to furan chemistry.

2-Methoxyfuran and 2-acetoxyfuran, which can be obtained by elimination from the corresponding 2,5-dihydro-2,5-dioxyfurans, undergo Diels Alder reactions and substitution just like other monosubstituted furans.

2 3-OXYFURANS

Here again the hydroxyl tautomer concentration is negligible in an equilibrium with the 3(2H)-furanone. These compounds are relatively unstable and resinify when left to stand. Alkali, in particular, causes rapid decomposition.

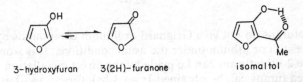

3-hydroxyfuran   3(2H)-furanone   isomaltol

The introduction of a carbonyl function at C 2 totally alters the situation, and makes the enol tautomer the dominant form (as part of a hydrogen bonded β-keto enol system).

3 2-AMINOFURANS

All attempts to prepare 2-aminofurans, including hydrolysis of the corresponding acetamide compounds or reduction of a nitro-precursor, do not lead to isolable products. It seems that simple 2-aminofurans are very unstable compounds. The only known examples of 2-aminofurans are those with electron withdrawing substituents, e.g., 3,5-dinitro-2-aminofuran.

2-Acylaminofurans are stable compounds, preparable by the reduction of a precursor-nitro-compound in the presence of acetic anhydride.

4 3-AMINOFURANS

3-Amino-2-methylfuran is a relatively stable compound which behaves

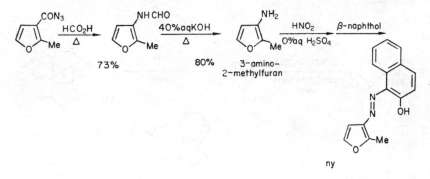

73%   80%   3-amino-
                    2-methylfuran

ny

as a normal aromatic amine and can be acylated and diazotized.

Nothing appears to be known about the position of the amino-imino tautomeric equilibrium.

M Synthesis of Furan Compounds

Furfural and, thence by vapour-phase decarbonylation, furan, are available in bulk and represent the starting points for many furan syntheses.

The aldehyde is manufactured from xylose, obtained in turn from pentosans, which are polysaccharides extracted from many plants, e.g. corn cobs and rice husks. Acid catalyses the overall loss of three moles of water from D-xylose in very good yield. The scheme gives the probable course of the reaction.

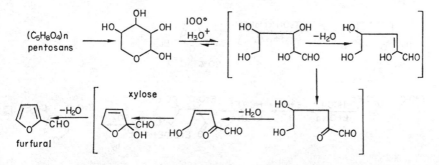

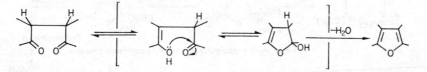

1 RING SYNTHESIS

There are two important general methods for the synthesis of furan rings from non-heterocyclic precursors, and these are set out schematically below.

(a) 1,4-Dicarbonyl compounds can be cyclized with loss of water.

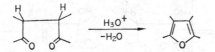

(b) α-Chlorocarbonyl compounds react with 1,3-dicarbonyl compounds in the presence of a base (not ammonia, see p. 215) as shown.

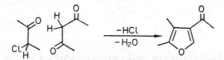

(a) From 1,4-dicarbonyl compounds

The Paal-Knorr Synthesis. The main limitation on the generality of this approach is the availability of an appropriate 1,4-dicarbonyl compound. Once obtained, the cyclization of such precursors, which provide

all of the carbon atoms and the oxygen atom necessary for the nucleus, generally proceeds in high yield. Usually, non-aqueous acidic conditions are employed to encourage the required loss of water. The process may involve the oxygen of the enol form of one of the carbonyl groups adding to the carbón of the other carbonyl group. Elimination of water then completes the process.

The examples shown give an idea of the scope of this synthetic route.

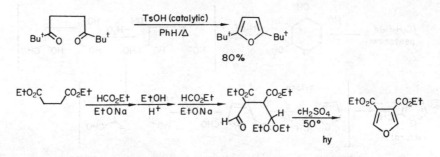

(b) From α-halocarbonyl compounds

The Feist-Benary Synthesis. In its simplest form, this synthesis probably involves an aldol condensation with the carbonyl group of the halogeno-component, followed by formation of the oxygen ring by intramolecular displacement of halide, and finally loss of water.

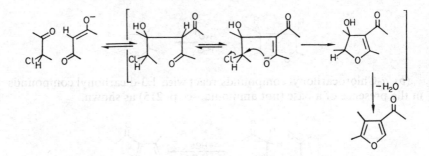

This synthesis nearly always proceeds, as shown in the scheme, through to an aromatic furan. In a few instances, however, the reaction stops short of a the final dehydrative step. For example, in the synthesis of furan tetracarboxylic ester, a 3-hydroxy-2,3-dihydrofuran is isolated. Although

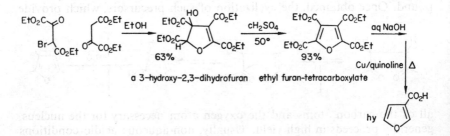

a 3–hydroxy–2,3–dihydrofuran   ethyl furan-tetracarboxylate

by no means a proof, the isolation of such a compound is a strong indication that the normal Feist-Benary synthesis proceeds *via* such a compound.
Decarboxylation of furan tetracarboxylic acid is the best way of preparing
β-furoic acid.

Chloroacetaldehyde leads to the formation of α,β-unsubstituted furans:
these may also be obtained by the use of α,β-dichloroethyl ether, as shown
in the following example.

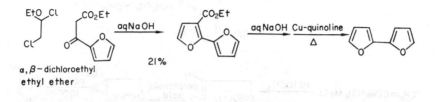

α,β-dichloroethyl
ethyl ether

It is very important to realize that 1,4-dicarbonyl compounds, such as
can be used in approach (*a*), are in fact often prepared by the alkylation of
a 1,3-dicarbonyl compound with an α-halocarbonyl compound. The
reaction of chloroacetone and acetoacetic ester shows how the same
materials can be used *either* to produce a 1,4-dicarbonyl compound *or*
directly in a Feist-Benary synthesis. Notice carefully that the two approaches give rise eventually to isomeric furans. In this case the aqueous

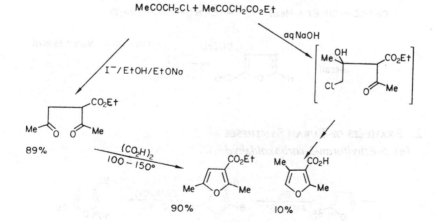

reaction medium favours aldol condensation as the first step, and thus a
furan is formed directly by the Feist-Benary approach. In alcohol solution,
however, the solvent properties favour displacement of halide as a first
step, and thus a 1,4-dicarbonyl compound is formed.

This can be cyclized to a furan by approach (*a*).

(*c*) *Miscellaneous methods.*   Many other methods have been used for the preparation of furan compounds, some of them have potential for general application. Given below are three examples of such approaches.

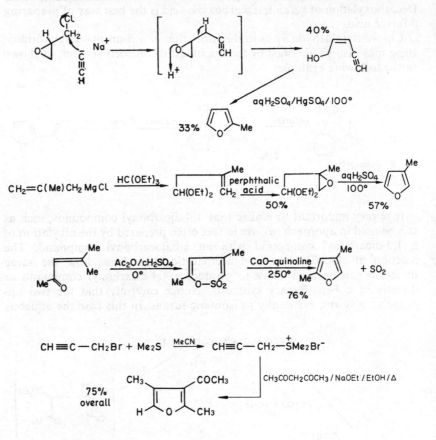

2   EXAMPLES OF FURAN SYNTHESES

(a) 5-Ethylfuran-3-carboxaldehyde

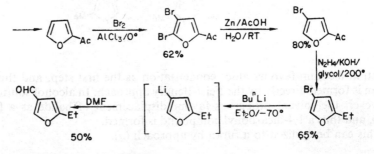

5-ethylfuran-3-carboxaldehyde

(b) *Iso-maltol* (isolated from bread; has a strong caramel flavour) and its conversion into *3-methoxyfuran*.

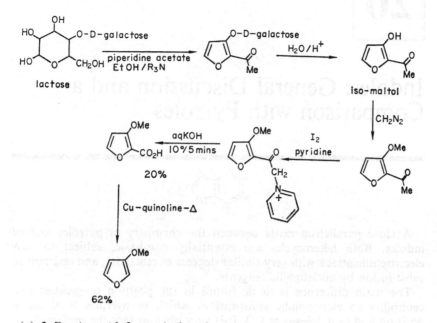

(c) *3-Furoic acid* from 4-phenyloxazole, an alternative method (cf. p. 252) illustrating the use of electrocyclic reactions.

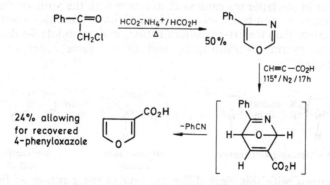

20

Indoles: General Discussion and a Comparison with Pyrroles

A close parallelism exists between the chemistry of pyrroles and of indoles. Both heterocycles are essentially non-basic, subject to easy electrophilic attack with very similar degrees of reactivity, and resistant to substitution by nucleophilic reagents.

The main difference is to be found in the position of greatest susceptibility to electrophilic substitution, which in pyrroles is at an α-position and in indoles is at C 3. It is very obvious that the preferred C 3 reactivity of indoles is the direct consequence of the presence of a benzene ring: this makes a 3H-indolium cation much stabler than a 2H-indolium cation, for in the latter the canonical structure with the positive charge on the nitrogen is non-benzenoid. A 3H-indolium cationic intermediate is also stabler than the 3H-pyrrolium cation, making indole β-substitution faster than pyrrole β-substitution, and of the same order as pyrrole α-substitution.

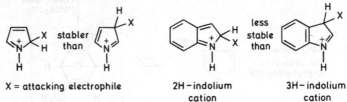

In contrast with this clear differentiation in the position of favoured electrophilic substitution, C-deprotonation of the N-alkylated heterocycles occurs at C 2 in both cases, since this involves the generation of a negative charge which is stabilized by *induction* by the electronegative nitrogen atom, and not by mesomeric delocalization, since it does not form part of the π-electronic system.

21

Indoles: Reactions and Synthesis

Indole and the simple alkyl indoles are colourless crystalline solids with a range of odours from naphthalene-like in the case of indole itself to faecal in the case of skatole (3-methylindole). Many simple indoles are available commercially and all of these are produced by synthesis: indole, for example, is made by the high temperature vapour-phase cyclizing dehydrogenation of *ortho*-ethylaniline, itself produced from *ortho*-nitroethylbenzene, a by-product of chloramphenicol synthesis.

Most indoles are quite stable in air with the exception of those which carry a simple alkyl group at C 2, thus 2-methylindole autoxidizes quite readily even in a dark brown bottle.

Indole was first prepared in 1866 by zinc dust distillation of oxindole. The word indole is derived from the word India: a blue dye imported from India became known as indigo in the sixteenth century, and chemical degradation of this gave rise to indoxyl, oxindole, and then to indole.

Tryptophan is an essential amino acid and as such is a constituent of most proteins. In animals, tryptophan also serves as a precursor for two chemically closely related hormones: serotonin (5-hydroxytryptamine) is a powerful vasoconstrictor and also a neurotransmitter substance and melatonin is believed to play a part in controlling the day and night rhythms of physiological functions. In the plant kingdom, β-indolylacetic acid is a plant growth-regulating hormone, also derived from tryptophan.

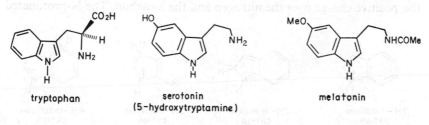

tryptophan

serotonin
(5-hydroxytryptamine)

melatonin

Many indolic secondary plant metabolites have potent physiological effects in man: a number of indole alkaloids have found widespread medical use, for example, reserpine as a tranquillizer, vincristine in the treatment of leukemia, ergotamine in the treatment of migraine, and ergonovine as an oxytocic agent. Among the synthetic chemotherapeutics, the β-indolyl-acetic acid derivative indomethacin is of value in the treatment of rheuma-toid arthritis. The activity of the semi-synthetic derivative LSD (lysergic acid diethylamide) is notoriously well known; the less familiar hallucinogen psilocybin occurs naturally in a Central American mushroom.

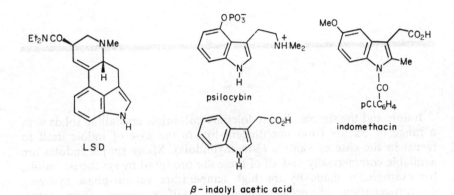

psilocybin

indomethacin

LSD

β - indolyl acetic acid

REACTIONS AND SYNTHESIS OF INDOLES

A   Reactions with Electrophilic Reagents

(a) *Protonation.*   Indoles, like pyrroles, are very weak bases; typical pK_a values are: indole, $-3\cdot5$, 3-methylindole, $-4\cdot6$, and 2-methylindole, $-0\cdot3$ This means, for example, that in 6M sulphuric acid two molecules of indole are protonated for every one unprotonated, whereas 2-methyl-lindole is almost completely protonated under the same conditions. Spectral data, mainly ultraviolet and NMR, show that in all cases in solution only the β-protonated cation, 3H-indolium, is detectable. The main reasons for its thermodynamic stability, and hence its dominance at equilibrium, are retention of benzene aromaticity and delocalization of the positive charge over the nitrogen and the α-carbon. The N-protonated

1H—indolium
cation

2H—indolium
cation

3H—indolium
cation

4H—indolium
cation

cation, 1H-indolium, likewise has an undisturbed benzene ring, the positive charge, however, is almost entirely localized on nitrogen: spectral data do not exclude its presence to the extent of 1 or 2 per cent. All the remaining seven possible cations have lost benzene aromaticity. It is important to note the close similarity between indoles and aliphatic enamines, most of which also are predominantly protonated at the β-carbon at equilibrium.

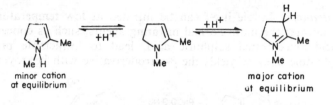

minor cation
at equilibrium

major cation
at equilibrium

Deuterium exchange studies show the N-hydrogen of indole to exchange most rapidly, indeed it exchanges at pH 7 when no exchange at all occurs at C 3 (see p. 290 for discussion of N-hydrogen exchange). With increasing acidity, increasingly rapid exchange at C 3 occurs by way of 3H-indolium cation and no exchange at C 2 or on the benzene ring occurs until relatively high acidity is reached, in other words acid-catalysed exchange at C 3 is very much faster than at C 2, C 4, C 5, C 6 or C 7.

That 2-methylindole is a much stronger base than indole is due mainly to stabilization of 2-methyl-3H-indolium cation by electron-release from

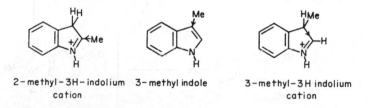

2−methyl−3H−indolium
cation

3−methyl indole

3−methyl−3H indolium
cation

the methyl group at C 2. That 3-methylindole is a weaker base than indole is probably due mainly to the fact that the methyl group can exert its conjugative stabilizing effect *only* in the unprotonated molecule; a second contributing factor may be that in the cation there is only one hydrogen at C 3 involved in hyperconjugation as opposed to two in the cation from indole itself.

Reactions of Protonated Indoles. In sharp contrast with indoles, 3H-indolium cations are electrophilic species and, under favourable equilibrium and solubility conditions, will react as such. Thus 3H-indolium cation is involved in the reaction with bisulphite to give the indoline 2-sulphonate salt under conditions which lead to its crystallization: it reverts to indole and sodium bisulphite on being dissolved in water.

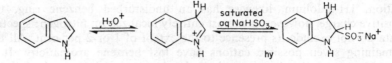

3H-indolium cations are also involved as electrophiles in the acid-catalysed dimerization and trimerization reactions of indoles (see pp. 267 and 276).

(*b*) *Nitration.* Indole itself can be nitrated at low temperature with benzoyl nitrate; the more normal nitrating agents, such as concentrated nitric acid-concentrated sulphuric acid, lead to intractable products. 2-Methylindole likewise yields the β-nitroderivative with benzoyl nitrate,

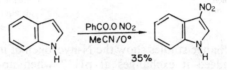

but because of its resistance to polymerization it also reacts smoothly with concentrated nitric acid or with concentrated nitric acid-concentrated

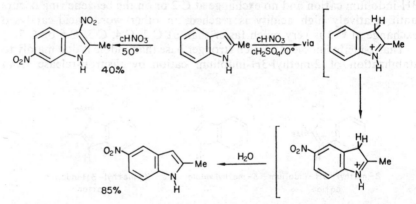

sulphuric acid mixture. The latter reagent gives an apparently anomalous reaction in which the β-position remains unsubstituted and C 5 is nitrated (see p. 268): this is explained by complete β-protonation of the 2-methylindole in the concentrated acid, so that nitration of the cation, and not of the free heterocycle, occurs. Predominant *meta*-nitration of aniline in concentrated sulphuric acid solution is an analogous phenomenon, which also involves nitration of a cation.

Nitration of skatole at C 2 has been achieved with benzoyl nitrate, but only in very low yield; the main reactions lead to complex intractable products.

(*c*) *Sulphonation* of indole occurs at C 3 using the pyridine sulphur trioxide compound in pyridine as solvent. Earlier claims to observation of

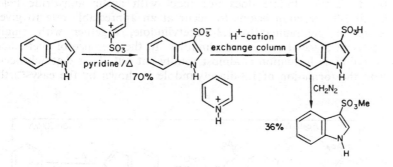

C 2-sulphonation with the same reagent under somewhat different conditions are probably in error.

(*d*) *Halogenation*. Very mild reagents must be used to effect C 3 halogenation of indole. In indoles where C 3 is already substituted, C 2

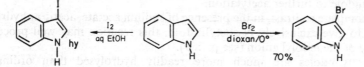

substitution occurs, but probably not in a straightforward manner.

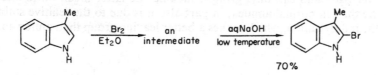

When skatole is treated with bromine in water, one of the main products is 5-bromo-3-methyloxindole.

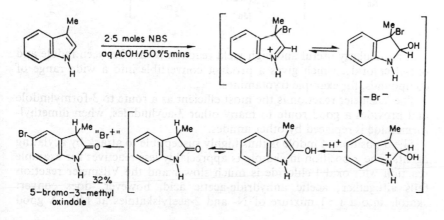

5-bromo-3-methyl
oxindole

(e) Acylation. Indole does not react with acetic anhydride below 100°; at 140° reaction begins to occur at an appreciable rate to give a mixture of predominantly 1,3-diacetylindole, together with smaller amounts of 3-acetyl and N-acetylindoles. In the presence of acetic acid, N-acetylindole formation is almost totally cut out. That β-attack occurs first in the formation of 1,3-diacetylindole is shown by the easy further

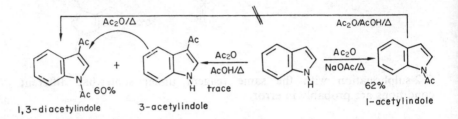

N-acetylation of 3-acetylindole as against the complete resistance of N-acetylindole to further acetylation.

By complete contrast, in the presence of sodium acetate, acetic anhydride reacts to give exclusively N-acetylindole: this reaction may well proceed by way of the indolyl anion (see p. 269).

N-acetylindoles are much more readily hydrolysed than ordinary amides, aqueous sodium hydroxide at room temperature being sufficient: this lability is in part due to the very much lower mesomeric interaction of the nitrogen and carbonyl groups, making the latter much more electrophilic than in normal amides; in part also it is due to the relative stability of the indolyl anion, which makes a better leaving group than amide anion.

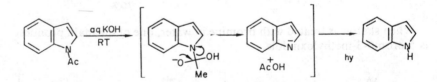

A particularly useful and high yield reaction is that between indole and oxalyl chloride, which gives a product convertible into a wide range of compounds, for example tryptamines.

The Vilsmeier reaction is the most efficient as a route to 3-formylindole and provides a good route to many other 3-acylindoles, when dimethylformamide is replaced by other amides.

The β-position of indole is thus highly susceptible to attack by acylating agents. The α-position in skatole is appreciably less reactive: for example reaction with oxalyl chloride is much slower and the Vilsmeier reaction fails altogether; acetic anhydride-acetic acid, however, does convert skatole into a 1 : 1 mixture of N- and 2-acetylskatoles at 140°. A good

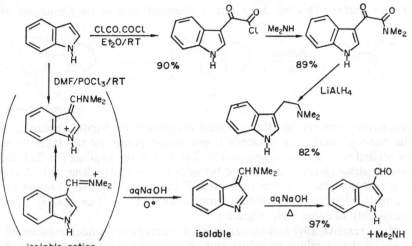

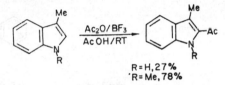

yield of 1-methyl-2-acetylskatole can be obtained by using a Lewis acid to activate the acylating agent.

$$R=H, 27\%$$
$$R=Me, 78\%$$

(*f*) *Alkylation*. (See also p. 273.) Indoles do not react with alkyl halides at room temperature. Indole itself begins to react with methyl iodide in DMF at 80–90°, when the main product is skatole. As the temperature is raised, further alkylation occurs until eventually 1,2,3,3-tetramethyl-3H-indolium iodide is formed.

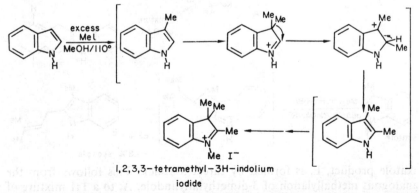

The rearrangement of 3,3-dialkyl-3H-indolium cations by alkyl migration to give 2,3-dialkylated indoles, such as that shown above, is related mechanistically to the Wagner-Meerwein rearrangement: thus substitution at indole C 2 *can* occur by alkylation at C 3 followed by rearrangement. This has been neatly demonstrated in the formation of

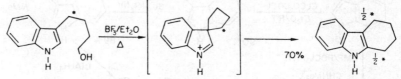

tetrahydrocarbazole

tetrahydrocarbazole by BF_3-catalysed cyclization of β-indolylbutanol: if the benzylic methylene is labelled, one might expect all the label to be associated with the methylene attached to C 3 in the product; in fact, the label is almost equally distributed between the two methylenes at C 2 and C 3, and this clearly proves that most of the reaction passes through the symmetrical spiro-3H-indolium cation in which there is an exactly equal probability of either CH_2 migrating.

More reactive allyl halides can effect C-substitution under mild conditions, as the reactions of indole and skatole with dimethylallyl bromide show: it is probable that an appreciable percentage of the 2-substituted

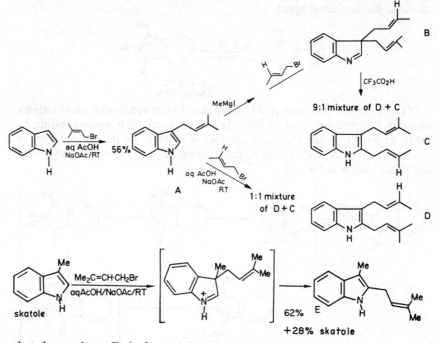

skatole product, E, is formed by direct α-attack. This follows from the analogous methallylation of 3-dimethylallylindole, A, to a 1:1 mixture of straightforward C and rearranged D products: that in this case about half the reaction proceeds by direct attack at C 2 is proved by the observation that acid-catalysed rearrangement of the 3H-indole, B, gives predominantly D (D:C = 9:1).

(g) *Condensation with aldehydes and ketones*. Indoles react with aldehydes and ketones by acid catalysis: the initial products, β-indolylcarbinols, are never isolated for they dehydrate to the 3-alkylidene-3H-indolium cations which, if sufficiently stabilized by mesomerism or by hyperconjugation, may be isolated: reaction with *p*-dimethylaminobenzaldehyde (the Ehrlich reaction, see p. 200) gives a mesomeric and highly-coloured cation; reaction of 2-methylindole with acetone under anhydrous conditions gives 3-isopropylidene-2-methyl-3H-indolium sulphate, the simplest isolable salt of this class.

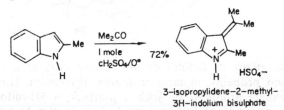

3-isopropylidene-2-methyl-
3H-indolium bisulphate

The β-alkylidene 3H-indolium cations are themselves usually good electrophiles and react with more of the indole, as is shown by the reaction with formaldehyde.

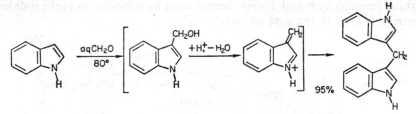

Skatole will also take part in this type of condensation at C 2.

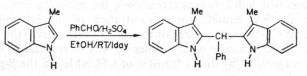

skatole

(h) *Alkylation by α,β-unsaturated carbonyl compounds*. This occurs by acid catalysis and may be looked upon as an extension of the reaction with ketones and aldehydes referred to above. Indole reacts with methyl vinyl ketone in acetic acid-acetic anhydride to give a β-alkylated product.

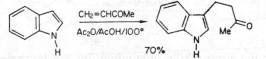

An instructive case is that provided by the reaction between 1,3-dimethylindole and mesityl oxide. This reaction illustrates the concept that

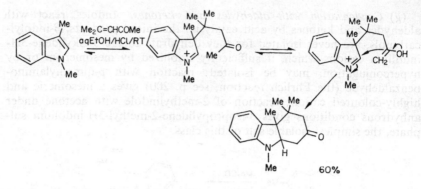

electrophilic attack of indoles occurs readily even at a *substituted β*-position (and probably, in most instances, reversibly). However, when the intermediate produced by such a *β*-attack, a 3H-indolium cation, also possesses a nucleophilic function which can cyclize by reaction at C 2, then the C 3-addition is secured even if it is usually reversible.

(*i*) *Condensation with immonium ions.*

The Mannich Reaction. Under neutral conditions and at low temperature, formaldehyde and dimethylamine react by substitution at the indole nitrogen, even in the case of indole itself:

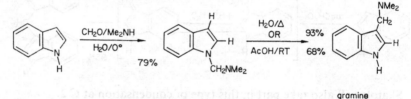

gramine

The reaction with indole is very remarkable and probably represents reaction with the low equilibrium concentration of indolyl anion: under more vigorous conditions at 100°, or in acid, conversion into the thermodynamically stabler gramine occurs either by way of reversal to indole and Mannich reagent, or by direct transfer of CH_2NMe_2 to the *β*-position of another molecule.

β-Substitution occurs smoothly and in high yield when the reaction is carried out in acetic acid. The Mannich reaction is a very useful reaction in

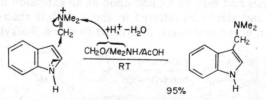

synthesis, for the electrophilic immonium ion can be varied widely, and gramine and related bases are important intermediates (see p. 275).

The acid catalysed dimerization of indole is an example of the Mannich reaction in which the 3H-indolium cation serves as the electrophilic

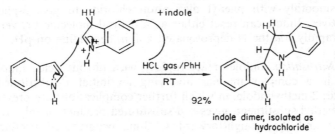

indole dimer, isolated as
hydrochloride

reagent for a second molecule of indole (see also pyrroles, p. 201).

Skatole reacts with formaldehyde and dimethylamine in acetic acid to give N-dimethylaminomethylskatole and not the 2-derivative, which is an indication of the relatively low reactivity of C 2. It is rather surprising, in view of this, to find that skatole dimerizes by what in effect is a Mannich substitution at C 2.

2-Methylindole does not dimerize, most probably because of the lower

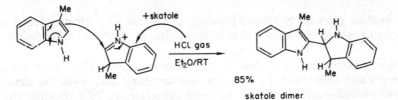

skatole dimer

electrophilic character of the corresponding cation combined with increased steric hindrance (cf. pyrroles, p. 202): these same factors render ketones less reactive than aldehydes.

Intramolecular Condensation at the α-Position. It must be appreciated that Vilsmeier and Mannich reactions leading to α-substitution *can occur* in intramolecular reactions where ring formation greatly increases the probability of reaction.

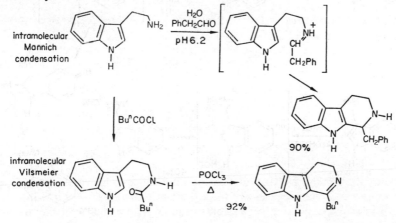

(*j*) *Diazocoupling.* Both indole and N-methylindole, but not skatole, react smoothly with phenyl diazonium chloride to give 3-phenylazo-derivatives. Indole can react either as a neutral molecule or, very much more rapidly, *via* the N-deprotonated anion, depending on pH.

(*k*) *Nitrosation.* Indoles react rapidly with nitrous acid; indole itself reacts in a complex manner involving an initial β-substituted inter-mediate; 2-methylindole, in which further complications are prevented by the 2-methyl substituent, gives a β-substituted product cleanly. Both this product and the straightforward C 3-monosubstituted product from indole, which *can* be obtained by reaction with amyl nitrite in the presence of base, exist largely as 3-oximino-3H-indole tautomers.

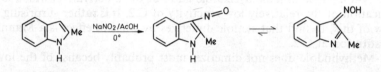

Skatole gives an N-nitroso-product, consistent with kinetic studies on 2-methylindole which show that N-substitution precedes C-substitution.

(*l*) *Substitution at carbon in the benzene ring.* This only occurs simply and cleanly either in concentrated sulphuric acid when the electro-phile probably attacks the 3H-indolium cation (see p. 260), or when a side chain positive charge sufficiently hinders β-addition of the electrophilic reagent: attack of the neutral indole nucleus then gives mainly 6-substitution, with some 4-substitution, as is seen in the nitration of gramine.

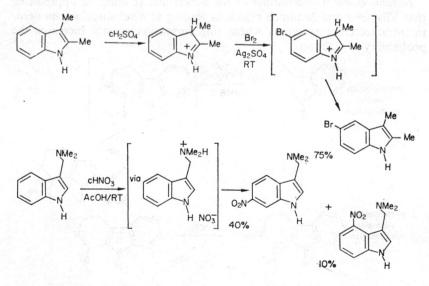

Reaction of simple 2,3-dialkylindoles in media more weakly acidic than concentrated sulphuric acid gives complex mixtures, almost certainly by way of initial β-addition of the electrophile: the bromination of skatole

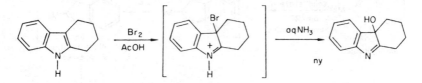

(p. 261) provides another example of this type of reaction.

(*m*) *N-Substitution.* (See also p. 273.) This has been observed in acylation, the Mannich reaction, and in nitrosation: here, one has to differentiate between reaction of the electrophilic reagent with a low equilibrium concentration of indolyl anion, which is probably occurring in acetylation in the presence of sodium acetate and in the Mannich reaction in the absence of acetic acid, and reaction with the actual neutral indole NH, which is probably occurring in the nitrosation of skatole.

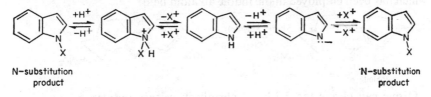

N—substitution   N—substitution
   product   product

B Reactions with Oxidizing Agents

Autoxidation occurs readily with alkyl indoles, thus for example 2,3-

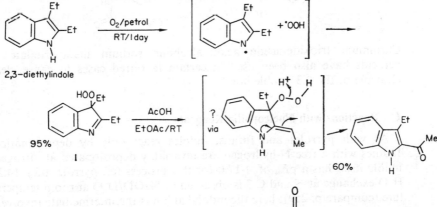

2,3—diethylindole

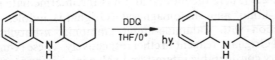

diethylindole gives an isolable 3-hydroperoxy-3H-indole, 1, as a product. The autoxidation of 2-methylindole is more complex, though still

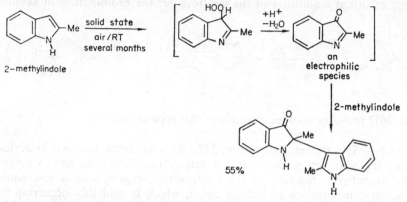

2-methylindole

an electrophilic species

2-methylindole

55%

initiated by oxidation of C 3. Selective oxidation of a β-CH$_2$ (and somewhat less efficiently of a β-methyl) can be effected with DDQ at 0°C.

Lead tetraacetate likewise reacts at the β-position: this is an oxidation which has been employed in the indole alkaloid field.

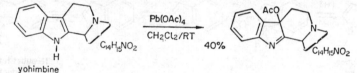

yohimbine

Ozone can cleave the 2,3-bond cleanly in certain indoles.

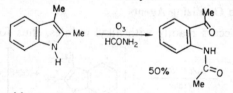

50%

Chromium trioxide-acetic acid, alcoholic sodium metaperiodate or peracids have also been used in certain favoured cases to cause clean cleavage of the 2,3-double bond.

C   Reactions with Nucleophilic Reagents

As with pyrroles and furans, indoles react only by deprotonation. Indoles with a free N-hydrogen are invariably deprotonated at nitrogen: indole itself has a pK_a of +17·0 for this process (cf. pyrrole, pK_a 14.2). H/D exchange at N and C 3 is observed in NaOD/D$_2$O at room temperature (compare p. 259); here the indolyl anion is the intermediate involved. The N-sodio-, -lithio and -halomagnesyl (Grignard) indoles are best formed by reaction with sodamide-liquid ammonia, n-butyllithium in ether and methyl Grignard reagent in ether respectively. These N-metallated indoles are of very considerable interest and value in synthesis (see p. 273).

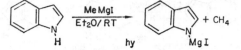

N-Alkylindoles are deprotonated at C 2 by butyllithium in exact analogy

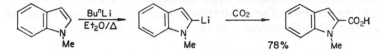

with the reactions of furans, thiophens and N-alkylpyrroles.

N-benzenesulphonylindoles can also be lithiated at C 2 and the resulting organometallic intermediates utilized synthetically, with final removal of the protecting group.

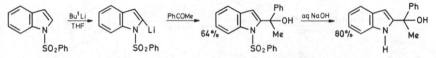

D   Reactions with Free Radicals

Relatively little work has been done on the reaction of indoles with free radicals. The thermal decomposition of *t*-butylperoxide in toluene is a method of generating benzyl radicals: if this is done in the presence of indole, then a very inefficient reaction occurs. In a typical experiment 50 per cent of the indole was recovered, 36 per cent was converted into intractable products, and 14 per cent was converted into a mixture of about equal proportions of 1-, 3-, and Bz-monobenzyl- and 1,3- and 2,3-dibenzylindoles.

On the other hand, when N-methylindole is allowed to react with benzoylperoxide in benzene at ambient temperature, smooth C 3 substitution by benzoyloxy occurs, though whether or not this is a free-radical reaction is not yet established.

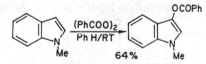

By complete contrast, reaction with hydroxyl radicals in water (H_2O_2-Fe(II)) leads to the isolation of products of reaction at the homocyclic positions only, giving the four phenolic indoles in poor yield.

E   Reactions with Reducing Agents

The neutral indole ring system is not reduced by nucleophilic reducing agents such as sodium-alcohol, lithium aluminium hydride or sodium borohydride. Lithium-liquid ammonia does however reduce the benzene ring.

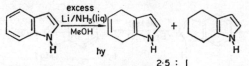

2·5 : 1

Reduction at C 2 and C 3 can be achieved with metal-acid combinations, but best with either sodium cyanoborohydride or triethylamineborane, both in the presence of acid. All these probably involve initial β-attack by an electrophile and reduction of the resulting 3H-indolium cation.

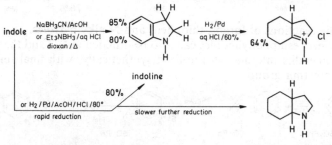

* note that aniline has been reduced to cyclohexanone under these conditions

Diborane effects reduction of indoles with a free NH.

Catalytic reduction can give indoline, or further reduction products depending on conditions, as can be seen from the scheme. Vigorous conditions can lead to rupture of the nitrogen-C 2 bond and formation of aniline derivatives.

F   Reactions with Dienophiles

Dienophiles do not react electrocyclically with indoles since this would necessarily disrupt benzene aromaticity. We have already seen how α,β-unsaturated carbonyl compounds react as simple electrophiles by acid catalysis (p. 265).

Benzyne has little affinity for indoles: N-methyl indole gives 4 per cent of 1-methyl-3-phenylindole; indole itself reacts as the anion to give a low

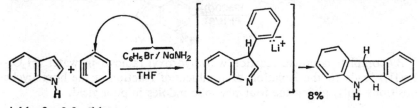

yield of a 2,3-adduct.

G   Reactions with Carbenes

2,3-Dimethylindole reacts with dichlorocarbene to give a mixture of 3-chloro-2,4-dimethylquinoline and 3-dichloromethyl-2,3-dimethyl-3H-indole.

Indole reacts in much the same way in a reaction which at one time was used to prepare 3-formylindole, 3-chloroquinoline here being the second

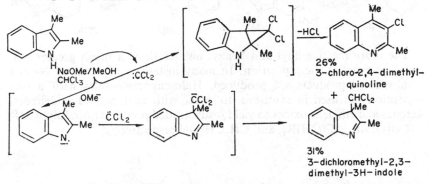

26%
3-chloro-2,4-dimethyl-quinoline

+

31%
3-dichloromethyl-2,3-dimethyl-3H-indole

product: the 3-formylindole must be produced by a pathway analogous to that which leads to the dichloromethyl-3H-indole in the example.

Carbethoxycarbene reacts with indole to give 3-indolylacetic ester:

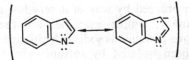

ethyl 3-indolylacetate

no cyclopropyl or ring-expanded products are isolated.

H   Reactions of N-Metallated Indoles

The indolyl anion is mesomeric (see p. 204) with negative charge delocalized, but mainly on the nitrogen and the β-carbon, as is seen in the two main canonical forms:

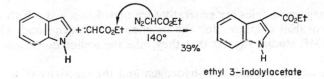

In its chemical reactions, therefore, this anion behaves as an ambident nucleophile, and the ratio N- to β- substitution by an electrophile depends on the nature of the associated metal cation, the substituents already present, the nature of the solvent, the temperature, and the nature of the electrophile. As a general rule, a sodio-indole reacts predominantly at the nitrogen, whereas a halomagnesylindole (indolyl Grignard) reacts mainly at the β-carbon. Several more convenient methods for N-alkylation, in some cases utilizing only an equilibrium concentration of the indolyl anion, are now available: combinations of indole, alkylating agent and NaH-HMPA, KOH-DMSO, 50%aqNaOH-PhH-Bu$_4$N$^+$HSO$_4^-$, or TlOEt-DMF all effect alkylation at room temperature and in very high yields.

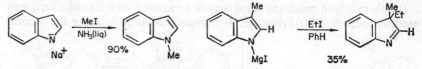

The more reactive allyl and benzyl halides show a much greater tendency to react at the β-position. In many instances, mixtures of N- and β-substituted products are produced. Halomagnesylindoles with a free β-position are used in synthesis to react with acyl halides, aldehydes, ketones and ethylene oxides to yield β-substituted products (β-indolyl-COR $-CH(OH)R, -C(OH)R_2$, and CH_2CH_2OH respectively).

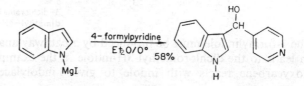

Halomagnesylindoles are generally formulated in a way which suggests a great covalent character for the magnesium-nitrogen bond. However, recent NMR studies suggest that they, like the sodioindoles, are ionic in character.

The acidity of the indole N-hydrogen and the reactivity of the indolyl anion are such that many reactions of indoles, carried out in aqueous

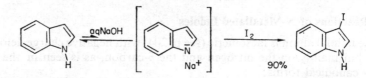

alkaline solution, can proceed by way of a very low equilibrium concentration of indolyl anion: simple examples are the β-iodination and β-chlorination of indoles in aqueous sodium hydroxide solution; N-chloroindole has recently been isolated by solvent extraction of an aqueous chlorination and has been shown to rearrange smoothly to 3-chloroindole, thus it seems that N-halogenation is kinetically favoured. The N-acetylation in the presence of sodium acetate (see p. 262) falls into the same general category. The N-nitrosation of skatole however occurs by attack on the neutral molecule.

I Alkylindoles

In the various substituted alkyl indoles, only alkyl groups at the α-position show any special reactions. Many observations indicate that the following equilibria occur in acid solution, thus the α- but not the β-methyl in 2,3-dimethylindole is deuterated in 3N deutero-hydrochloric

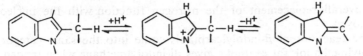

acid at 100°. Evidence from chemical reaction comes from the Mannich reaction of 1,2,3-trimethyl indole, when condensation takes place, *via*

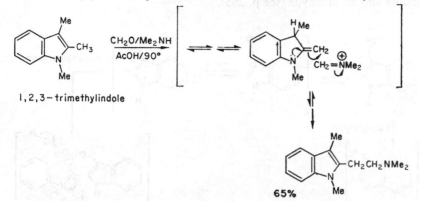

1,2,3−trimethylindole

65%

an enamine intermediate, at the 2-methyl group. In a similar way, the C 2 methyl of 1,2-dimethylindole has been oximated with nitrous acid and that of 2,3-dimethylindole coupled with a reactive diazonium salt.

J   Reactions of Indole-C-X Compounds

Gramine and, especially, its quaternary salts are extremely useful synthetic intermediates, in that they are easily prepared and in that the nitrogen group is easily displaced by nucleophiles. The following two reactions are typical:

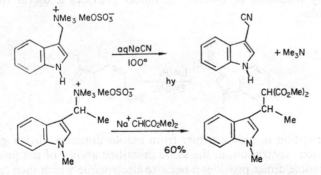

60%

The easy displacement of the nitrogen group from gramine salts is believed to proceed by ionization of the carbon-nitrogen bond which leads to highly reactive β-methylene indolenine, or indoleninium species. These conjugated imine or immonium compounds are very susceptible to nucleophilic addition at the terminal carbon, which process brings about the overall replacement of the nitrogen function with the nucleophile employed.

The reactions of β-indolylcarbinols come into the same category: β-indolylmethanol, for example, gives di-indolylmethane when treated with acid. A reasonable sequence of events is pictured. Notice that the final step in the sequence implies the reverse of a condensation between an indole and an aldehyde (see p. 265).

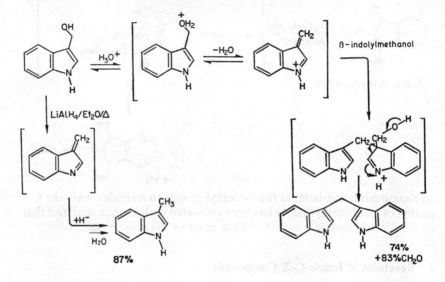

It is of interest to note that whereas β-indolylmethanol is reduced to skatole by lithium aluminium hydride, the N-methyl-analogue is not: this shows the reaction not to be a direct displacement of the oxygen function.

This easy reduction of β-indolylcarbinols provides a useful route to β-alkylindoles.

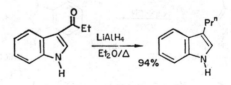

The formation of indole trimer from indole dimer is also a gramine type reaction: ionization, in the sense described above, of the protonated form of indole dimer provides a reactive electrophile which then combines with a third indole molecule (reaction sequence on next page).

K Indolecarboxylic Acids

Both indole 3-carboxylic acids and indolyl α-acetic acids are easily

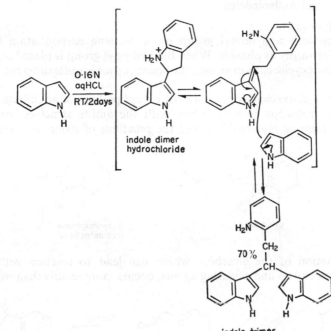

decarboxylated in boiling water. Almost certainly the carbon dioxide is

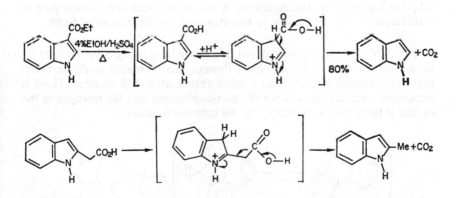

lost from a low equilibrium concentration of a β-protonated species. It is noteworthy that in the 3H-indolium cation the β-aromatic carbon and α-alkyl positions are equivalent and that both the above cations are structurally analogous to β-keto acids, so that the easy loss of carbon dioxide is readily understood. Indole 2-carboxylic acids also lose carbon dioxide easily on heating in mineral acid (see p. 287).

L   Oxy- and Aminoindoles

1   Oxyindoles

Indoles with a hydroxyl group on a benzene carbon atom behave straightforwardly as phenols. When the hydroxyl group is placed on either of the heterocyclic carbon atoms, the consequence is quite different.

(*a*)   *2-Hydroxyindole* does not exist except perhaps at an infinitesimally small concentration in equilibrium with the amide structure, *oxindole*. There is nothing remarkable about the reactions of oxindole except that

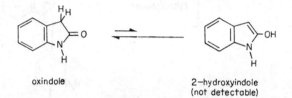

oxindole

2–hydroxyindole
(not detectable)

deprotonation of the β-carbon, which can lead to reaction with alkyl halides, aldehydes and acylating agents, occurs more readily than with, say

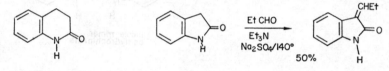

3,4,–dihydro–2–quinolone

50%

dihydro-2-quinolone: this reactivity is associated with the extra degree of stabilization of the anion due to the aromatic indolic canonical form.

(*b*)   *3-Hydroxyindole*, or indoxyl, is a more interesting case, for here we have a very reactive molecule which, though it exists mainly as the carbonyl tautomer, almost certainly has a quite appreciable enol content. There is mesomeric interaction between the carbonyl group and the nitrogen in the carbonyl tautomer, as indicated by the canonical forms.

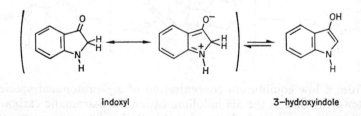

indoxyl   3–hydroxyindole

Deprotonation (of either tautomer) occurs readily to give an ambident mesomeric anion, on which electrophilic attack can occur at either oxygen or carbon depending on the precise conditions.

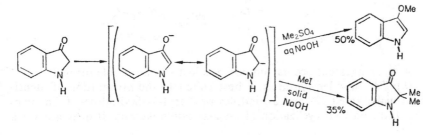

Indoxyl is particularly easily autoxidized to indigo by a free radical mechanism involving loss of an electron from the anion as a first step.

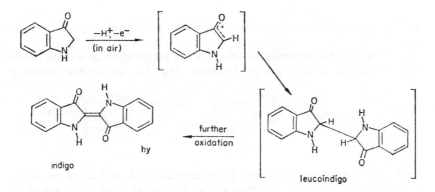

2   AMINOINDOLES

Whilst little is known of the structure of 3-aminoindole, the 2-isomer exists mainly as the 3-H-indole amino-tautomer, probably because of stabilization by the amidine-type mesomerism, 2, as shown.

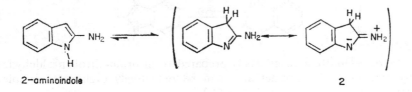

M   Synthesis of Indole Compounds

1   RING SYNTHESIS

There are five main general routes for the synthesis of the indole ring system from non-heterocyclic precursors; naturally, all of them begin with benzene derivatives.

(*a*) The Fischer synthesis is by far the most widely used indole synthesis. It consists of heating a phenyl hydrazone, most often with acid, though sometimes in an inert solvent alone.

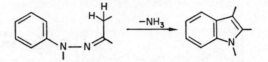

A rearrangement occurs, ammonia is lost and an indole formed.

(*b*) The Bischler synthesis is best suited to the preparation of identically substituted (2- and 3-) indoles or 2-aryl-3-alkylindoles. It involves the acid-catalysed cyclization of an α-arylaminoketone. It does not work

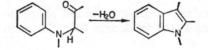

with benzene compounds having an electron-withdrawing group on the ring.

(*c*) The Reissert synthesis is useful for the preparation of indoles carrying various substituents on the benzene ring. An *ortho*-nitrotoluene is

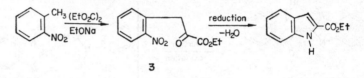

3

condensed with oxalate to give an intermediate, 3, which on reduction cyclizes to give an indole-2-carboxylic acid, or ester. Such acids are easily decarboxylated.

(*d*) The Madelung synthesis yields 2-alkyl and 2-arylindoles. The synthesis involves the very vigorous base treatment of an *ortho*-toluidide. The vigour of the reaction conditions restricts its applicability.

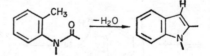

(*e*) *o*, ω-Dinitrostyrenes, easily prepared from *ortho*-nitrobenzaldehydes by reaction with nitromethane, can be reductively cyclized to indoles carrying no substituents at C 2 or C 3.

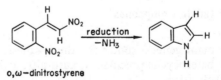

o,ω−dinitrostyrene

(*a*) *From phenylhydrazones of aldehydes or ketones.*

The Fischer Synthesis. This involves the acid-catalysed rearrangement of a phenylhydrazone with the elimination of ammonia. In many instances this reaction can quite simply be carried out by mixing aldehyde or ketone with one mole equivalent of phenylhydrazine in acetic acid, and refluxing. The formation of the phenylhydrazone and its subsequent rearrangement thus take place without the isolation of the intermediate. The preparation of skatole and 2-methylindole illustrate the synthesis in its simplest form.

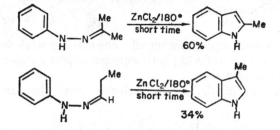

Indole cannot be obtained from acetaldehyde phenylhydrazone under the usual liquid phase reaction conditions, but this has now been achieved in the vapour phase at 300° over zinc chloride.

The full details of the mechanism of this multi-step reaction have not been worked out, but there is considerable evidence to suggest the following sequence.

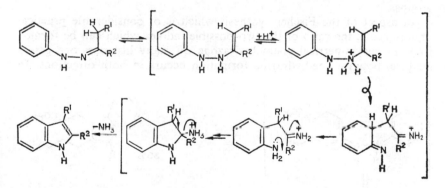

The most important step is that in which a carbon-carbon bond is made. This step is very likely to be electrocyclic in character and thus analogous to the Claisen rearrangement of phenyl allyl ethers.

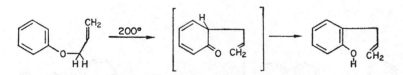

Support for this view comes from the observation that in many cases the Fischer synthesis may be achieved simply by heating a phenyl-hydrazone to 200° in the absence of acid. Recently, indolization has been achieved thermally at a temperature as low as 110° in the special case of ene-hydrazines: the first step of the normal sequence has already been achieved for these compounds.

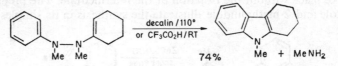

The reaction does occur more rapidly, however, by weak acid catalysis, in which proton adds to the aliphatic nitrogen to give a cation which then undergoes rearrangement.

Acid treatment of the ene-hydrazine 4 gives the salt 5: in this

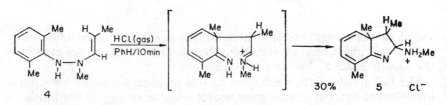

particular instance aromatization is prevented by the *ortho*-methyl groups.

An aspect of the Fischer synthesis which is of considerable practical importance is the ratio of the two possible indoles which can be formed from an unsymmetrical phenylhydrazone. In many instances one gets a mixture, because ene-hydrazine formation occurs in both directions. In

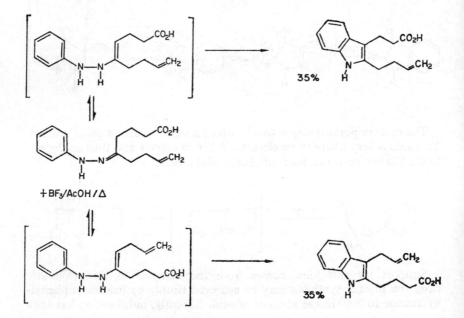

other cases one ene-hydrazine, and hence one derived indole, is strongly favoured. It has recently been discovered that the direction of closure can be controlled by choice of acid and solvent. A fully satisfactory rationalization of such cases has yet to be found.

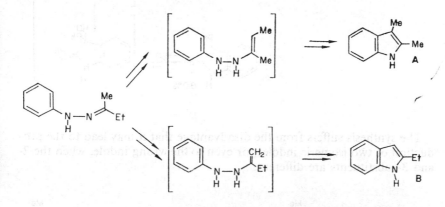

reaction in 66% aqH₂SO₄ leads mainly to A, whereas in 86% H₂SO₄ it
leads to a 1:1 mixture of A and B

Indolenines (3H-indoles) are frequently products of the Fischer reaction using branched ketones, the example shown illustrates this.

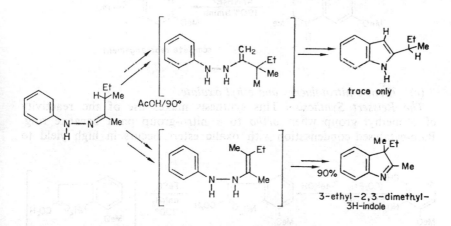

(b)   *From arylamines and α-halocarbonyl compounds.*

The Bischler Synthesis. This is the acid- or zinc chloride-catalysed dehydrative cyclization of α-arylaminoketones, which are generally easily prepared from 2-haloketones and an arylamine. The cyclizing step involves electrophilic attack of the benzene ring by protonated (or zinc chloride complexed) carbonyl.

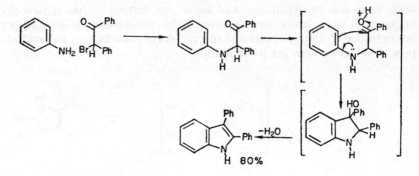

The synthesis suffers from the disadvantage that it may lead to the production of two isomeric indoles, or even to the wrong indole, when the 2- and 3-substituents are different.

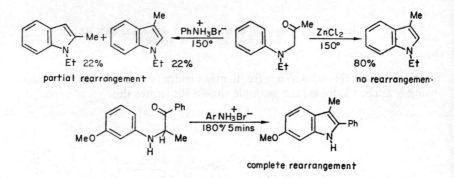

(c) From o-nitrotoluenes and ethyl oxalate.

The Reissert Synthesis. This synthesis makes use of the reactivity of a methyl group when *ortho* to a nitro-group on a benzene ring. Base-catalysed condensation with oxalic esters occurs in high yield to

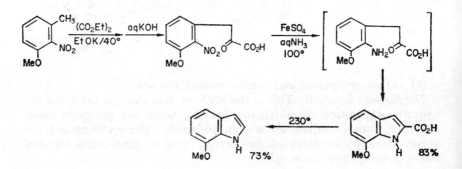

give intermediates which are converted into indole 2-carboxylic acids by mild reduction of the nitro-group with ferrous hydroxide, or into indole 2-esters by catalytic reduction. This approach has been much used for the preparation by decarboxylation of 2,3-unsubstituted indoles, which in turn are important starting points. The usefulness of the procedure is governed mainly by the availability of the substituted o-nitrotoluene.

(d) From o-toluidides.

The Madelung Synthesis has the advantage of starting with simple materials, ortho-toluidines and an appropriate carboxylic acid. The process consists of the base-catalysed (best with $NaNH_2$ or $NaOBu^t$) cyclization of an ortho-toluidide. The full mechanism is not known, but must involve deprotonation of the aryl-methyl group at some stage, and nucleophilic addition to the amide carbon.

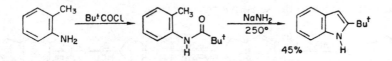

This reaction is used mainly for the synthesis of simple 2-alkyl and 2-arylindoles as well as 2,3-unsubstituted indoles carrying alkyl or aryl substituents on the benzene ring.

A new, mechanistically similar process, by which indole itself or 3-alkyl-2-unsubstituted indoles can be synthesised, also depends on the formation of aryl-methylene or aryl-methine anions and their intramolecular addition, in this case to an isocyanide function.

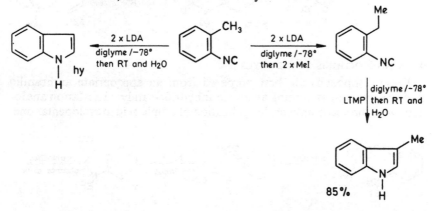

(e) From o, ω-dinitrostyrenes.

The reductive cyclization of o,ω-dinitrostyrenes provides a good route to many benzene-substituted 2,3-unsubstituted indoles; its value depends on the ease with which the appropriately substituted ortho-nitrobenzaldehyde can be synthesized.

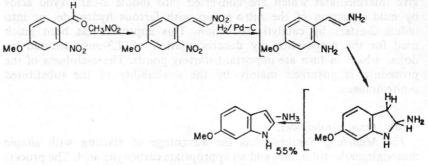

2 Synthesis of Indoles from Indolines

Of the many other methods of synthesis, one which has been much used more recently for 2,3-unsubstituted indoles involves the intermediacy of indolines, 2,3-dihydroindoles, with dehydrogenation to the indole as the last step. This allows transformations to the benzene ring to be carried out without risk to the sensitive pyrrole ring.

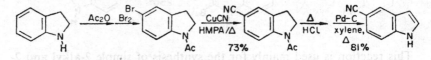

3 Ring Synthesis of Oxindoles

The main synthesis is simple and direct and involves a Friedel-Crafts reaction as the cyclizing step.

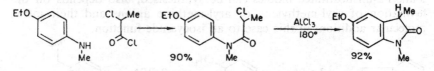

4 Ring Synthesis of Indoxyls

These compounds are best prepared from an appropriate anthranilic acid, the cyclizing step being an acetic anhydride-catalysed acylation analogous with the acetic anhydride cyclization of adipic acid to cyclopentanone.

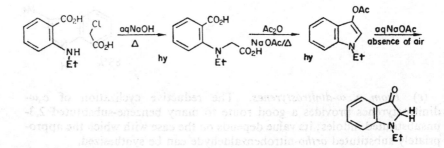

5 EXAMPLES OF INDOLE SYNTHESES

(a) *Serotonin* is a naturally occurring vasoconstrictor, it may play an important role in the operation of the central nervous system.

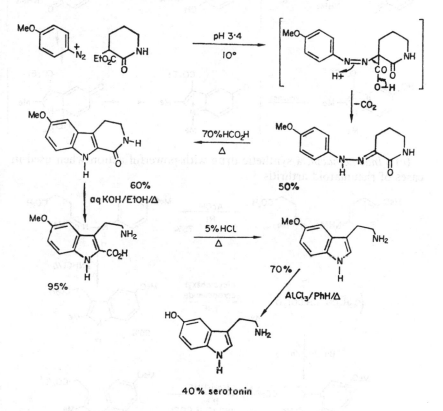

40% serotonin

(b) *5-Hydroxyindoles*. The Nenitzescu synthesis is a general method for 5-hydroxyindoles, a class of compounds in which there is considerable biological interest (see p. 257). Precise details of the oxidation-reduction processes which must occur during such syntheses are not known with certainty, a reasonable sequence is outlined in the example below.

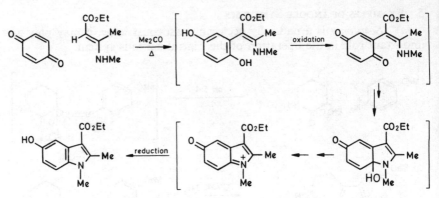

(c) *Indomethacin*, a synthetic drug with powerful action when used in cases of rheumatoid arthritis.

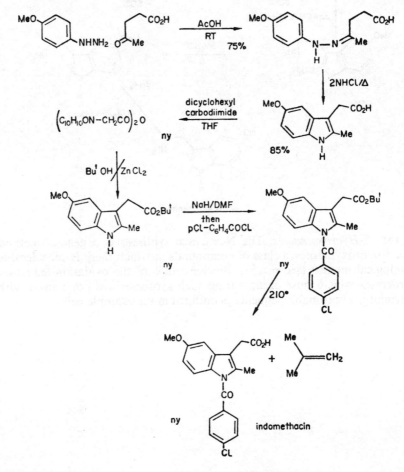

(*d*) *Tryptophan*, an essential amino acid.

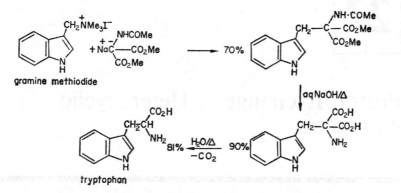

gramine methiodide

tryptophan

22

Proton Exchange at Heterocyclic Nitrogen

Proton Exchange at N in Amides, Amines, Pyrroles, Indoles, Pyridones and Related Compounds

Hydrogen exchange at N occurs very rapidly in these compounds at pH's around neutrality, a phenomenon which is of value in NMR spectroscopy, when an NH (or OH) signal can be removed by the simple addition of a drop of D_2O to the solution in the tube. This easy exchange is at first sight surprising, for around neutrality the equilibrium concentration of protonated or deprotonated species is extremely low, especially with amides and the pyrroles and indoles: the main reason for the rapid exchange is that the activation energy for NH or OH bond-making or breaking is very low, and this makes such reactions very favoured kinetically. Thus, even though the actual equilibrium concentration of the protonated or deprotonated species 1–4 may be very low, they are being formed very frequently and their low concentration is the consequence of their very short life.

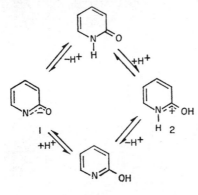

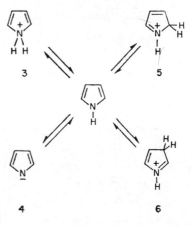

Protonation of Pyrroles and Indoles at Carbon

Over a relatively wide range of pH around neutrality, where NH exchange occurs rapidly, CH exchange does not occur at an appreciable rate. It might seem surprising that at lower pH values, which allow an equilibrium concentration of a protonated species to be observed spectroscopically, it is found that this species is C-protonated, being mainly cation 5 in the case of pyrrole. One may understand this apparent anomaly with reference to the discussion above. The argument here is that even though cation 5 formed by C 2 protonation of pyrrole is thermodynamically more stable than cation 3 (see p. 185), the activation energy for CH bond formation to give 5 is very much higher than that for NH bond formation to give 3. This means that the rate of formation of 5 is very much slower than that of 3, so that at neutral pH, formation of 5 and exchange at carbon is very slow. It is only when a low pH forces the build up of a spectrally detectable quantity of C-protonated cation, that exchange at carbon becomes rapid. Even though cation 3 must inevitably be formed even faster in acid solution, the low activation energy for the reverse process and the instability of cation 3 with respect to neutral pyrrole, mean that its equilibrium concentration remains low.

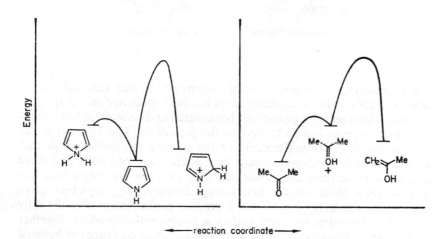

reaction coordinate

Closely related to this is the extremely slow deuteration of acetone in D_2O, even though the equilibrium concentration of acetone enol is appreciable, being about one molecule for every 50,000 keto: here again, C protonation-deprotonation are the slow steps.

Benzofurans and Benzothiophens: General Discussion and a Comparison with Indoles

benzo[b]thiophen   benzo[b]furan

The reactivity of indole towards electrophiles and the high regio-selectivity for C 3 in such substitutions has been discussed earlier (p. 256) and might lead one to suppose that benzofuran and benzothiophen should also react more or less exclusively at the β-position. This however is not so: various kinetic measurements of the relative reactivities of β- and α-positions of benzothiophen have shown ratios between unity and three and, furthermore, that the homocyclic ring positions are only a little less reactive than those on the hetero-ring. Preparatively, acylation gives mixtures in which the β-substitution product predominates over the α by about 2:1. Nitration too gives mainly a 3-substitution product, together with appreciable quantities of the 2-isomer and small quantities of benzene ring substituted products. Controlled halogenation follows the same pattern in giving 3-halobenzothiophens as major products, however 2,3-disubstitution is difficult to prevent. In benzofuran the trend away from exclusive β-substitution is carried even further: only α-substitution products have been obtained from nitration and Vilsmeier formylation; Friedel-Crafts acetylation yields a mixture composed of 83 per cent 2-acetyl and 17 per cent 3-acetylbenzofuran. Halogenation is complex, with addition being the preferred mode, however even here, as can be seen in the analogous, clean addition of bromine azide, the electrophilic portion of the reagent becomes attached to an α-position, giving in this instance 3-azido-2-bromo-2,3-dihydrobenzofuran as product.

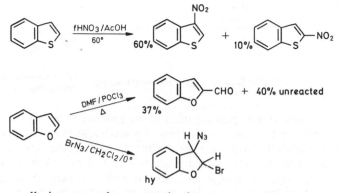

The overall picture can be summarized:

	α-substitution	β-substitution
indole	negligible	almost exclusive
benzothiophen	appreciable	dominant
benzofuran	dominant	appreciable

A comparison of the rates of electrophilic substitution in the pairs furan/benzofuran and thiophen/benzothiophen shows that the effect of the annelation of a ring is to reduce the overall reactivity. However, a comparison of partial rate factors for halogenation and acylation for the α- and β-positions in the two pairs shows that this overall reduction is the net effect of a large depression in the rate of α-substitution set against an increase in the rate of β-substitution. Why annelation should have this effect has already been argued for indole (p. 256); the same argument applies to benzofuran and benzothiophen.

Thus the observed orientations of substitution in benzofuran and benzothiophen are seen as reflecting the greater proportion of β-monosubstitution in thiophen compared with β-monosubstitution in furan. An extrapolation of these concepts includes pyrrole and indole, for pyrrole gives very considerable proportions of β-monosubstitution, nearly 20 per cent in the case of nitration, which then becomes almost exclusive β-substitution in indole.

More difficult to explain is why β-monosubstitution is negligible in furan, occurs to a minor extent in thiophen, and is quite appreciable in pyrrole. One of the key factors here must be the electronegativity of the heteroatom: the strongly electronegative character of oxygen cuts down the extent to which an unshared pair interacts with the two double bonds to form the aromatic system, so that furan reacts to a considerable extent as a conjugated diene in which one or other of the terminal positions, C 2 or C 5, would be expected to add an electrophile almost exclusively. This is analogous with, say, butadiene to which proton adds exclusively terminally to give the allyl carbonium ion.

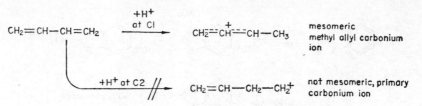

mesomeric methyl allyl carbonium ion

not mesomeric, primary carbonium ion

The very much greater participation of the unshared electrons of nitrogen and sulphur in the aromatic systems of pyrrole and thiophen make for an electron availability more evenly distributed over the four carbon atoms. In a reaction intermediate involving addition to C 3 or C 4 the canonical structure with positive charge on nitrogen makes a very much greater contribution than the corresponding canonical structure in an intermediate for C 3 substitution in furan.

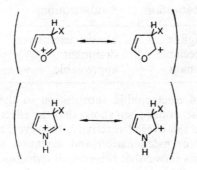

lesser contribution by the canonical stucture with + ve charge on O

very much greater contribution by canonical structure with + ve charge on N

Isoindoles and Indolizines:
General Discussion and a Comparison
with Indoles

Indole, isoindole, and indolizine are very similar both in ring structure and in being 10-electron aromatic systems: all three have a partially positive nitrogen, and all react easily with electrophilic reagents at positions 1 and 3.

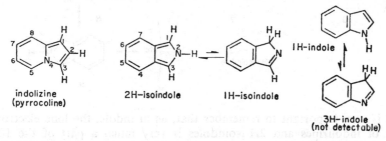

indolizine
(pyrrocoline)   2H−isoindole   1H−isoindole   1H-indole

3H−indole
(not detectable)

Indolizine is the only system which is not potentially tautomeric; though indole is potentially tautomeric, 3H-indole is so much less stable than the normal 1H-indole that it has never been observed; isoindole is potentially tautomeric but appears to exist largely in the fully aromatic 2H-tautomeric form in hexane or acetone solution. 1-Phenylisoindole has been shown to contain 10 per cent of the 3H-tautomer in $CDCl_3$ solution; the less conjugated 1H-tautomer could not be detected. Thus it seems that the 10-electron aromaticity of 2H-isoindoles is sufficient to hold such structures largely in what might have seemed to be the less favoured quinoid structure.

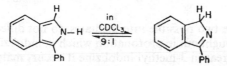

1−phenyl−2H−isoindole   1−phenyl−3H−isoindole

The mesomerism of indolizine is expressed by the main canonical structures 1, 2 and 3: structures such as 4 and 5 are probably not very im-

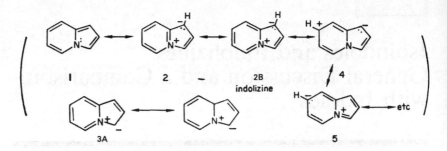

portant. The mesomerism of 2H-isoindole is best expressed by canonical structures 6, 7 and 8, with structures such as 9 contributing but little.

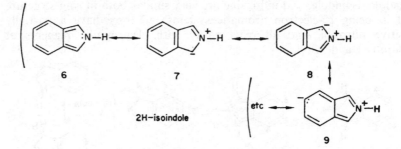

It is very important to remember that, as in indole, the lone electron pair of indolizines and 2H-isoindoles is very much a part of the 10-electron aromatic system.

In reaction with electrophiles, both for isoindole and for indolizine, the stablest intermediates 10, 11 and 12 involve addition to C 1 and C 3, in close analogy with corresponding intermediate in indole chemistry.

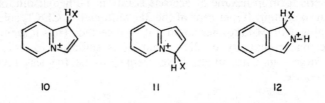

In indolizines the two possible intermediates do not differ very much in stability: this is brought out in protonation which in indolizine itself occurs mainly at C 3, whereas in 3-methyl indolizine it occurs mainly at C 1. The pK_a values show indolizines to be very much more strongly basic than indoles. The greater stability of indolizinium cations makes them relatively weak electrophiles, as is seen in the resistance of indolizine to acid-catalysed polymerization.

Indolizine	2-methyl-	3-methyl-	1,2-dimethyl-	2,3-dimethyl-
pK_a 3·9	5·9	3·9	7·3	5·4

Indole	2-methyl-	3-methyl-
pK_a −3·5	−0·3	−4·6

The pK_a's of isoindole or of its simple derivatives are not known. Protonation almost certainly occurs at C 1 or C 3; in contrast with indolizinium cations, isoindolium cations are probably quite electrophilic, in keeping with the rapid resinification of 2-methylisoindole in dilute acid.

Both 2H-isoindoles and indolizines react very readily by substitution at C 1 and C 3 with a wide range of electrophiles: indolizines are, broadly speaking, of the same order of reactivity as indoles, whereas 2H-isoindoles probably are more reactive, as evidenced by the very mild conditions required to effect C-acetylation of 1-phenylisoindole.

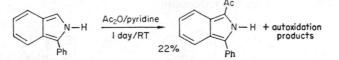

Indolizines, being stable compounds, have been much more extensively studied than isoindoles: they undergo easy alkylation, Vilsmeier formylation, diazocoupling, nitration, etc., at C 1 and C 3.

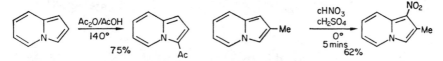

Isoindole itself can be isolated as a white solid at very low temperature; decomposition occurs if the solid is brought to room temperature. A special feature of the chemistry of all isoindoles is a great tendency to undergo Diels-Alder addition at C 1 and C 3, much like furans and a clean

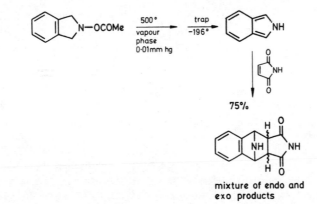

mixture of endo and
exo products

reaction between isoindole itself and a dienophile can be effected in solution.

Oxidation of isoindoles readily gives phthalic acid derivatives, and oxidation of indolizines gives picolinic acid derivatives, thus as in indoles the five-membered ring is the more susceptible to attack.

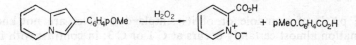

Reduction of isoindoles, whether by dissolving metals or by a catalytic process, likewise occurs in the five-membered ring to give a benzene derivative: on the other hand, catalytic hydrogenation of indolizines in neutral solution occurs in the six-membered ring to give a pyrrole derivative. Of some interest is the alternative catalytic hydrogenation of indolizinium cations using Pd-C to give the pyridinium compound, 13; platinum in acid solution catalyses full reduction.

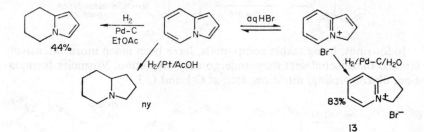

25

1,3-Azoles: General Discussion and a Comparison with Pyrrole, Thiophen and Furan and also with Pyridine

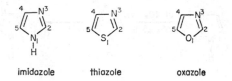

imidazole   thiazole   oxazole

Pyridine is included in this discussion because the 1,3-azoles are related to pyrrole, furan and thiophen much in the same way as pyridine is related to benzene.

Electrophilic Addition to N 3

Electrophiles can add to the N 3 (the azomethine nitrogen) of 1,3-azoles just as they can add to the nitrogen of pyridine, since the azomethine lone pair of electrons is not involved in the aromatic sextet.

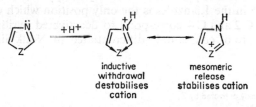

inductive   mesomeric
withdrawal   release
destabilises   stabilises cation
cation

The basicity of the azomethine nitrogen of 1,3-azoles is profoundly affected by interaction with the heteroatom X: in oxazole, the oxygen exerts a powerful inductive withdrawal and only a relatively weak mesomeric release, so the result is a much more weakly basic nitrogen than that of pyridine; in imidazole the inductive withdrawal by N 1 is weak, but the mesomeric release very strong, so imidazole is a much stronger base than pyridine.

Electrophilic C-Substitution

REACTIVITY

The 1,3-azoles are considerably less susceptible to electrophilic substitution than are the monohetero analogues, pyrrole, furan and thiophen,
and the reasons parallel those advanced in explanation of the lower reactivity of pyridine when compared with benzene (p. 34).

Electron-withdrawal in free base both makes the approach by the
electrophile more difficult and destabilizes the transition state for substitution. Very little precise information is available on electrophilic substitution of 1,3-azoles, thus in most cases it is not known whether reaction
proceeds by way of free base or conjugate acid.

Imidazole is clearly the most reactive of the trio, just as pyrrole is more
reactive than furan or thiophen; being the most basic, however, imidazole
is the most likely to have to react by its conjugate acid. In one of the few
carefully studied reactions in this category it has been shown that nitration

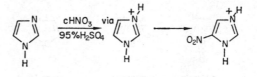

of imidazole in 95 per cent sulphuric acid proceeds by way of the salt (for
imidazolium cation mesomerism, see p. 304) and at a rate 10^{-9} that of the
nitration of benzene. Pyrrole is, of course, much more reactive than
benzene; pyridine, on the other hand, when reacting as the conjugate acid,
nitrates at about 10^{-10} the rate of imidazole.

POSITION OF ELECTROPHILIC SUBSTITUTION

Electrophilic substitution in the 1,3-azoles occurs mainly at C 5, which
makes sense in terms of pyridine β-reactivity and pyrrole-furan-thiophen α-
reactivity: C 5 in the 1,3-azoles is the only position which corresponds to
both whereas C 2 and C 4 correspond to deactivated pyridine α-positions
(C 4 and C 5 are of course only differentiated in N-alkylimidazoles, see
p. 305).

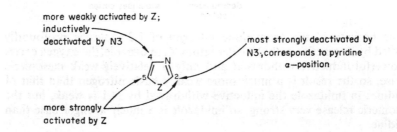

The mesomeric aspects of this analysis are better seen by considering the main canonical forms of the intermediates of reaction at the three different positions:

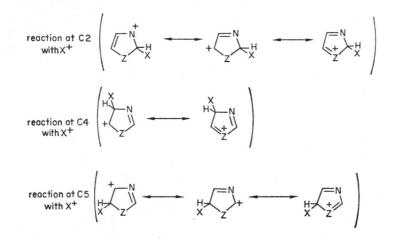

Reaction at C 2 is seen to involve a canonical form with a highly unfavoured sextet positive N 3.

Reaction with Nucleophiles

Nucleophilic substitution with displacement of H in 1,3-azoles has only been observed in a few instances in thiazoles when it occurs at C 2: oxazoles and imidazoles do not react in this manner. This makes a marked contrast with pyridines.

The 1,3-azoles, however, react quite readily by ring-proton abstraction, most particularly from C 2: in non-hydroxylic solvents a very strong base like butyllithium leads to complete C 2 deprotonation, thus N-benzylimidazole gives the 2-lithio derivative; reversible deprotonation by a weaker base such as MeO⁻ in MeOH is observable at room temperature with oxazoles and thiazoles and at higher temperatures with imidazoles, the equilibrium concentration of the heterocyclic anions being very low; the most interesting deprotonating situation is observable with all three 1,3-azoles in neutral methanol, when the low equilibrium concentration of the 3-protonated azolium cation is reversibly deprotonated at C 2 to give an ylid, 1, R = H, the equilibrium concentration of which is very low. Analogous ylids, 1, R = Alk., are involved in the C 2 exchange of quaternary salts of all three systems.

It is of interest to note that the 1H-pyridinium cation is also deprotonated at C 2 to give a transient ylid, even if under much more vigorous conditions; pyridine is also deprotonated at all positions in aqueous sodium hydroxide, again under vigorous conditions (see p. 57). Furans, thiophen and N-alkyl pyrroles are only deprotonated by the strongest bases, but then again specifically at C 2. The behaviour of the 1,3-azoles can thus be seen as a combination of the properties in the two simpler systems.

26

1,3-Azoles: Reactions and Synthesis

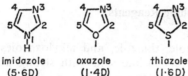

imidazole oxazole thiazole
(5·6D) (1·4D) (1·6D) dipole moment

The three 1,3-azoles, imidazole, oxazole, and thiazole are all very stable compounds which do not autoxidize. Oxazole and thiazole are water-miscible liquids with pyridine-like odours and normal boiling points, 69° and 117° respectively. Imidazole and 1-methylimidazole are water soluble and odourless, however they have very much higher boiling points, 256° and 199° respectively, which is probably due to dipolar association (a result of the very marked permanent charge separation between the two nitrogens, which is much greater than that in oxazole or thiazole as is evident in their dipole moments), and in addition in the case of imidazole itself to extensive hydrogen bonding.

Oxazole is the only one of the trio which does not play any part in fundamental metabolism: the imidazole system occurs in the essential α-aminoacid histidine, which plays a part in enzyme catalysis of hydrolysis

histidine   histamine

(see p. 305), and in the related hormone histamine, one of whose functions is also connected with digestion; the thiazole ring is the chemically active centre in the important co-enzyme thiamin (see p. 313).

The oxazole ring system is not involved in any chemotherapeutically useful compound, whereas the thiazole system is, as for example in succinoyl sulphathiazole, one of many sulphonamide antibacterial drugs, and so is the imidazole system, as in metromidazole which is active against a very wide range of micro-organisms, its most important use being in the treatment of amoebic dysentery.

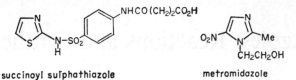

succinoyl sulphathiazole   metromidazole

REACTIONS AND SYNTHESIS OF THE 1, 3-AZOLES— IMIDAZOLE, THIAZOLE AND OXAZOLE

A   Reactions with Electrophilic Reagents

1   ADDITION TO NITROGEN

(a) *Protonation.* Imidazole, thiazole, and alkyloxazoles (though not oxazole itself) form stable crystalline salts with strong protic acids by protonation at the azomethine nitrogen, N 3.

Imidazole, with a pK_a of 7·1, is a very much stronger base than thiazole (pK_a 2·5) and oxazole (pK_a 0·8), and is even stronger than pyridine (pK_a 5.2). This is almost certainly due to a combination of the relatively low electronegative character of N and of the symmetrical structure of the mesomeric imidazolium cation, which bears a close resemblance to an amidinium cation. The particularly weakly-basic character of oxazoles is

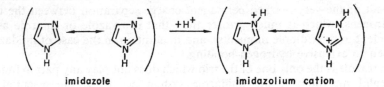

imidazole   imidazolium cation

principally due to the strong inductive effect of the oxygen towards itself, coupled with the weaker mesomeric electron release from it.

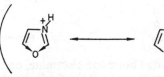

base weakening   much less favoured
inductive effect   canonical structure

The 1,3-azoles are quite stable even in strong acid under vigorous conditions.

Hydrogen-bonding in imidazoles. Imidazole is a very good hydrogen-bonding molecule and resembles water in that it is both a very good donor and a very good acceptor. This property of imidazoles is put to good use in nature by hydrolytic enzymes, in which histidine plays a central role as

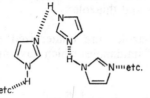

H− Bonded oligomeric imidazole

outlined in the scheme shown: this is believed to represent the mode of

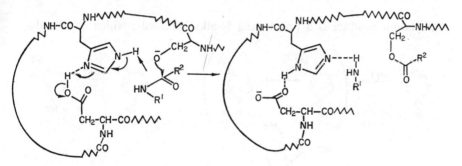

operation of the digestive enzyme chymotrypsin in the hydrolysis of amide links in proteins and peptides.

Tautomerism in imidazoles. Imidazoles with a ring NH are subject to tautomerism which becomes evident in unsymmetrically substituted compounds such as the methylimidazole shown. This special feature of imidazole chemistry means that to write 4-methylimidazole would be

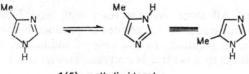

4(5)−methylimidazole

misleading, for this molecule is in tautomeric equilibrium with 5-methylimidazole, and quite inseparable from it. All such tautomeric pairs are inseparable, and the convention used to cover this phenomenon is to write 4(5)-methylimidazole.

In some cases, such as that of 4(5)-nitroimidazole, one tautomer has been shown to predominate greatly.

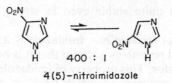

4(5)-nitroimidazole

Of course, tautomerism of this kind is not possible for N-substituted imidazoles, nor for oxazoles and thiazoles.

(b) *Alkylation.* The 1,3-azoles are quaternized easily at N 3 with alkyl halides. In the case of imidazoles which have an NH, the resulting

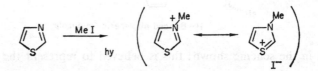

alkylation product is a protonated N-alkylimidazole, which can be de-

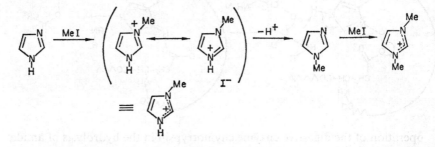

protonated by unreacted imidazole, and alkylated to 1,3-dialkylimidazolium salt: in practice then, imidazole gives a mixture of imidazolium, 1-methylimidazolium, and 1,3-dimethylimidazolium salts. Furthermore, an unsymmetrically-substituted imidazole will give two isomeric 1-alkyl derivatives.

(c) *Acylation.* All three systems react with acyl halides and, to a lesser extent, with carboxylic anhydrides to produce highly reactive N 3-acyl-onium salts in solution. In the case of imidazole, of course, N 1 deprotonation leads to a neutral N-acetylimidazole: this is a very reactive

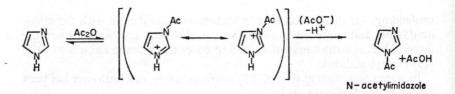

N-acetylimidazole

amide, and is easily hydrolysed to the parent heterocycle, even more easily than is N-acetylpyrrole. The way in which N-acetylimidazole can serve as a specific N-acylating agent for pyrrole (see p. 197) is not understood.

N-acyl imidazoles react with LAH to give aldehydes in a preparatively useful reaction.

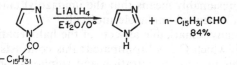

2   SUBSTITUTION AT CARBON

Much still needs to be clarified with the electrophilic substitution reactions of the 1,3-azoles, although it is quite clear that they are intermediate in reactivity between pyridine on the one hand, and pyrrole, furan, and thiophen on the other.

(a) *Protonation.* N-Methylimidazole is deuterated by DCl-D_2O at 160° by way of non-aromatic di-cations, such as that shown below: the relative rates at C 2, C 4, and C 5 are 1:73:120, in marked contrast with the almost exclusive C 2 deuteration by the ylid mechanism (see p. 311).

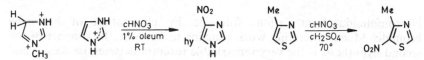

(b) *Nitration.* Imidazole is much more susceptible to electrophilic substitution than is thiazole. Kinetic studies have shown that nitration of imidazole involves electrophilic attack on the imidazolium cation, and not the neutral molecule.

Thiazole is much less reactive, for it is not nitrated in oleum at 160°: 4-methylthiazole, however, is nitrated at C 5 under relatively mild conditions. The relative rates of nitration of 4-methylthiazole, benzene, and thiophen are $1 : 110 : 3\cdot5 \times 10^6$.

It is important to note that C 2 in both heterocycles is very resistant to nitration, thus 4,5-dimethylimidazole does not react even under vigorous nitrating conditions.

(c) *Sulphonation.* Here again thiazole is seen to be much less reactive.

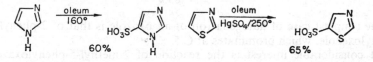

(d) *Halogenation.* The imidazoles are brominated with remarkable ease, N-methylimidazole, for example, reacts with one mole of bromine in water at RT to give 2,4,5-tribromoimidazole as main product (with about 70 per cent of unchanged N-methylimidazole). Chlorine and iodine, however,

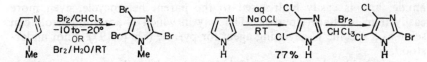

appear to react only with imidazoles containing NH, and only in alkaline solution: this presumably means that the imidazolyl anion is the reactive substrate, not the neutral imidazole molecule.

One of the many remarkable aspects of the halogenation of imidazoles is the ease with which C 2 is brominated: this contrasts sharply with the resistance of this position to nitration and sulphonation. The reason for this difference has not been established: it may be that bromination involves electrophilic attack of the neutral imidazole molecule, and not of the imidazolium cation as is the proven case in nitration, or it may be that C 2 substitution proceeds by nucleophilic addition of bromide to the

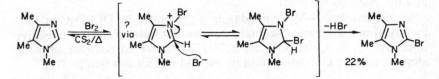

N-bromoimidazolium cation, followed by elimination of hydrogen bromide. Much interesting work remains to be done. In support of the second hypothesis is the very remarkable reaction between imidazole and cyanogen bromide, which leads to 2-bromoimidazole and hydrogen cyanide.

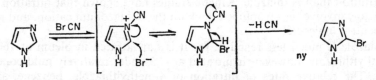

Thiazole does not react with bromine in chloroform, conditions which effect C 5 bromination of 2,4-dimethylthiazole.

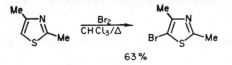

63%

The simplest case of halogenation of an oxazole is that of 2-phenyl-4-methyloxazole, which brominates at C 5.

Of considerable interest is the reaction of 2-methyl-5-phenyloxazole, isomeric with the compound mentioned above, with bromine in methanol: as the reactive C 5 is already substituted, simple substitution cannot take place, but a reaction occurs which probably involves addition of Br+ to C 5 as the first step, and which closely parallels the corresponding reaction of furan (see p. 242).

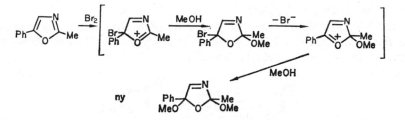

(e) Acylation. Though Friedel-Crafts acylation has not been observed, C 2 acylation of N-alkylimidazoles with aroyl halides and trihaloacetyl chlorides in the presence of triethylamine occurs readily. The mechanism is believed to involve C 2 deprotonation of an intermediate N-acyl imidazolium cation. This acylation fails with simple aliphatic acyl halides and with N-unsubstituted imidazoles.

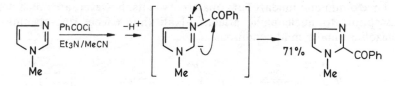

B   Reactions with Oxidizing Agents

All three ring systems are resistant to mild oxidation, but little is known of the degree of resistance relative to each other and to other ring systems.

Thiazoles give N-oxides with peroxidic reagents, but imidazoles break down. Oxygen in the presence of light and a sensitizer reacts to give a variety of products depending on the ring system, but in each case the primary product is likely to be a bridged peroxy adduct as in the corresponding reactions of furans and pyrroles.

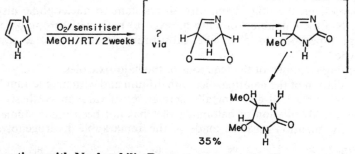

C   Reactions with Nucleophilic Reagents

Broadly speaking, the 1,3-azoles do not react by nucleophilic substitution: the few reactions in this category are at the C 2 of thiazole, which here again demonstrates its closeness to pyridine. On the other hand, C-deprotonation, especially at C 2, provides some very interesting chemistry.

1   SUBSTITUTION AT CARBON

(*a*) *With hydride transfer*. The amination of thiazole is the only example of this type of reaction in this group of heterocycles. N-alkylimidazoles are inert to sodamide in dimethylaniline at 125°.

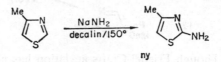

ny

Oxazoles, thiazoles, and imidazoles are stable to alkalis under vigorous conditions, for example 2-methyl-4-phenyloxazole is recovered unchanged from exposure to potassium hydroxide in ethanol at 200°.

The strongest bases, such as butyllithium, do not react with thiazole by addition, as they do with pyridine, but by proton abstraction (see p. 312).

Thiazolium and imidazolium quaternary salts, however, are much more susceptible to nucleophilic attack: alkali thus effects ring-opening of thiazolium under mild conditions.

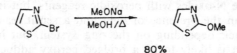

(*b*) *With displacement of halide*. 2-, 4-, and 5-halothiazoles all react easily with methoxide by nucleophilic displacement of halide, a 5-halogen

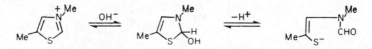

80%

being displaced most readily, and a 4-halogen least readily. The halo-imidazoles, on the other hand, are all resistant to nucleophilic displacement: N-alkyl haloimidazoles are believed to react with strong bases like lithium piperidide to give cine substitution products *via* a transient hetaryne.

Nothing is known of the reactions of the halo-oxazoles.

Metallation of the halothiazoles with lithium and with magnesium by the entrainment method gives metallo derivatives of value in synthesis in the usual way. Metallation of haloimidazoles has not been much studied.

Finally, mention must be made of the remarkable hydrogenolysis of

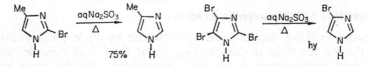

bromo- and iodo-imidazoles by hot aqueous sodium sulphite: this reaction occurs most readily at C 2, but also at C 4(5). The mechanism is believed to proceed by bromonium (or iodonium) abstraction by $SO_3^=$ to generate an intermediate C 2 ylid or, much more slowly, a C 4 ylid.

2   DEPROTONATION

(a) *Deprotonation of imidazole NH.* The pK_a for loss of the N-hydrogen of imidazole is 14·2: it is thus a very weak acid, appreciably stronger however than pyrrole (pK_a 17·5), because of the enhanced

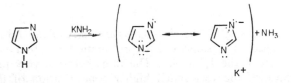

delocalization possible in the imidazolyl anion. A strong base, like sodium amide, is required to effect complete deprotonation.

(b) *Deprotonation of CH in azoles.*   The three heterocyclic systems undergo H-D exchange in neutral O-deuteromethanol. Since the rate of exchange falls to zero in acid, the reaction is not of the electrophilic displacement type, involving initial addition of proton (see p. 307) to carbon.

For imidazole, for which exchange under these conditions is about ten times faster than for oxazole or thiazole, the mechanism certainly involves N 3 protonation followed by C 2 deprotonation by methoxide anion to

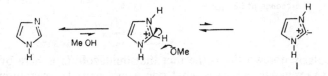

give a short-lived ylid, 1, which then accepts a deuteron from solvent in a fast reaction, thus effecting exchange. Although oxazole and thiazole can, and certainly do, undergo exchange by this type of mechanism, their much lower basicities and the consequent much lower concentrations of salt present in methanol solution make it likely that their exchange proceeds principally by C 2 proton abstraction directly from the uncharged molecule.

The oxygen and sulphur systems are in fact much better able to support a negative charge at C 2; thus, oxazole exchanges at this position in MeOH-MeONa at room temperature, conditions which do not affect N-benzylimidazole.

Under strongly-basic conditions all three systems can exchange by direct proton abstraction, not only at C 2 but at C 5 as well.

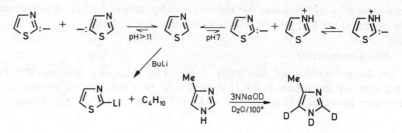

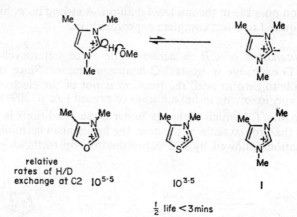

(c) *Deprotonation of C 2 in the quaternary salts of the 1,3-azoles.* As one would expect, the same C 2 deprotonation is observed with the quaternary salts of the 1,3-azoles. The rates here are very much faster, largely because of the very much higher concentration of the relevant azolium cations, even in alkaline pH's.

relative rates of H/D exchange at C2　　$10^{5.5}$　　　$10^{3.5}$　　　I

$\frac{1}{2}$ life < 3 mins

Particularly noteworthy is the fact that imidazolium is here by far the slowest to exchange, which is what one would expect: oxazolium, which contains the strongly-electronegative oxygen best able to enhance the acidity of the C 2 hydrogen, is much the fastest.

This special aspect of the reactivity of the 1,3-azoles is also exploited by the life process, for the catalytic action of the coenzyme thiamin is believed to involve a thiazolium ylid: the first step in its role as a catalyst for the decarboxylation of pyruvate is the addition of the thiamin ylid, 2, to the ketonic carbonyl of the pyruvate unit.

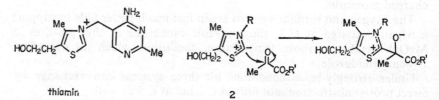

thiamin　　　　　　　　　　　　　　　　　　　　　　2

(d) *Reactions involving N- and ring C-deprotonated species.* In appropriately buffered solution, imidazole reacts with relatively weak electrophiles. Such reactions proceed by electrophilic attack on low equilibrium concentrations of N-deprotonated imidazolyl anion, 3. Iodination at pH 7 or above gives 4,5-di-iodoimidazole. Diazo-coupling, on the other

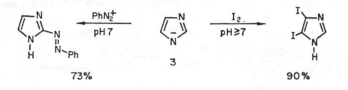

73%   90%

hand, occurs at C 2: no explanation for this remarkable difference is evident. Neither reaction occurs with N-methylimidazole.

Mannich reaction takes place at N 3 in acid solution, however in basic media reaction at all carbon positions occurs *via* the imidazolyl anion.

A reaction which very probably involves the intermediacy of a C 2 ylid is the very rapid methanolysis of 2-benzoyl-3,4-dimethylthiazolium iodide to methyl benzoate and 3,4-dimethylthiazolium iodide.

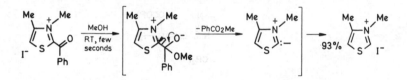

It is tempting to place speculatively in this section the very anomalous but very efficient reaction of imidazoles with formaldehyde in neutral solution to give 2-hydroxymethyl derivatives. The reaction could proceed

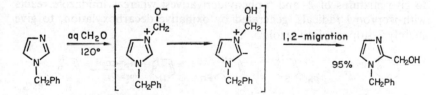

as shown, or more simply by electrophilic addition of CH_2O to the 3-NH ylid. A close study of this reaction would be rewarding.

(e) *Reactions involving side-chain deprotonation.* A methyl at C 2 in the three 1,3-azoles undergoes deprotonation under various conditions: butyllithium deprotonates the neutral molecule, whereas weak bases can deprotonate positively charged rings.

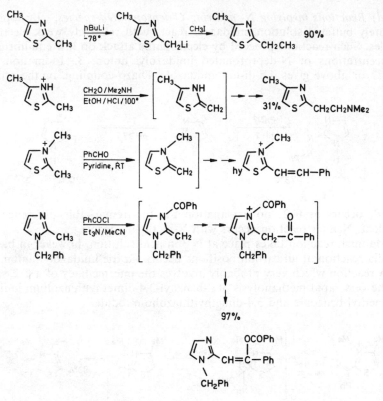

D   Reactions with Radical Reagents

The limited results available in this area suggest that the preferred points of attack are C 2 and C 5, much as in pyrrole and furan, with a marked preference in some cases for C 2.

Thiazole and 1-methylimidazole react with phenyl radicals in solution to give mixtures of 2- and 5-phenylderivatives, whereas imidazole reacts with isopropyl radicals, generated by oxidative decarboxylation, to give mainly 2-isopropylimidazole.

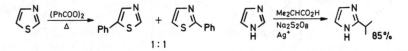

E   Reactions with Reducing Agents

Imidazole is the most resistant to reduction, thus it is unaffected by sodium-liquid ammonia, concentrated hydriodic acid and red phosphorus, zinc and acid, or H_2 and a catalyst. Thiazole is also resistant, but a very active Raney nickel catalyst will desulphurize it. Oxazoles seem to

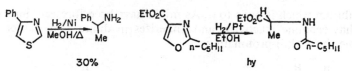

30%   hy

be the most reactive, for whilst resistant to zinc and acetic, they are ring-opened with C-O fission by Na-EtOH or by Pt-H$_2$.

F   Reactions with Dienophiles

Here a contrast emerges between oxazole on the one hand and thiazole and imidazole on the other, which is paralleled in the chemistry of furan, pyrrole, and thiophen. Standard Diels-Alder addition reactions occur

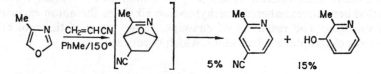

with oxazole and a variety of substituted oxazoles, reactions which have considerable utility in synthesis as is seen in a recent pyridoxine synthesis (see pp. 320 and 255). Thiazoles and imidazoles, however, react by nucleophilic addition to the dienophiles, as does pyridine, to give initially

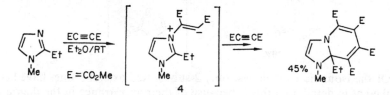

an N 3-quaternary salt, 4, which eventually cyclizes by nucleophilic addition at C 2.

G   1,3-Azolones

All the compounds in this group which have been studied exist as the carbonyl tautomers, which is in keeping with what we have learned so far about these tautomeric equilibria. Much work remains to be done, especially on the chemical reactivity of these systems.

In keeping with the carbonyl structure, the imidazol-4(5)-one, 5, is resolvable, though nothing is known of the position of the tautomeric

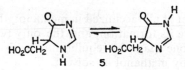

equilibrium which is open to it. As might have been foreseen, imidazol-2-one has enamine reactivity, and dimerizes in dilute acid, rather like a pyrrole, by C 4(5)-protonation.

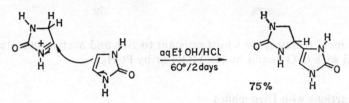

75%

The three thiazolones are known, and the only reaction carried out on all of them so far has been the standard conversion into the corresponding halothiazole by phosphoryl halide. Most work has been done on the thiazol-2-ones: diazomethane reacts to give a mixture of O-methyl and N-methyl derivatives, whereas methyl iodide alkylates the anion exclusively on nitrogen. The reactivity of enamine, rather than of thioenol ether, controls electrophilic substitution, which occurs at C 5.

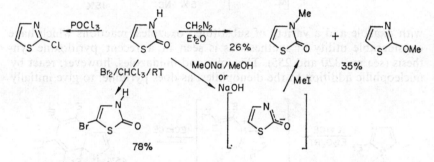

Of the isomeric oxazolones, only 2-substituted oxazol-5-ones have been looked at in detail and this is because of their appearance in the degradation of penicillin. One of the simplest is 4-isobutyl-2-phenyloxazol-5-one which exists exclusively as the carbonyl tautomer in ether (as is seen in the retention of optical activity when it is synthesized from benzoyl L-leucine) but in equilibrium with an appreciable proportion of hydroxy tautomer in methanol on the following argument: the optically active oxazolone reacts with methanol to give racemic benzoyl leucine; now since optically-active

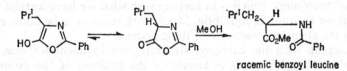

racemic benzoyl leucine

benzoyl leucine is not at all racemized in methanol, the racemization must have occurred before its formation, and the only possibility is racemization at the oxazolone stage by equilibrium with the achiral hydroxy tautomer favoured by methanol as solvent.

Oxazol-5-ones condense easily at C 4 by base catalysis with carbonyl compounds to give azlactones: this condensation is the first step in the well-

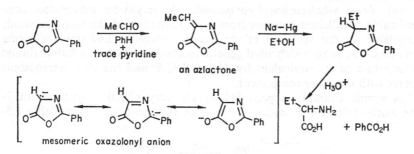

known azlactone synthesis of α-aminoacids, and may involve generation of a reactive mesomeric oxazolonyl anion by C 4 deprotonation.

H Amino-1,3-Azoles

All the known compounds in this class exist as amino tautomers, and undergo all the normal arylamine reactions including diazotization.

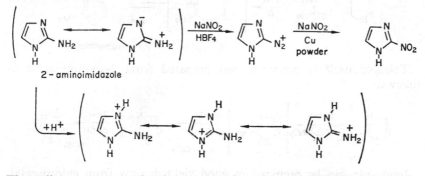

They all protonate on the ring azomethine nitrogen, N 3. Given that the 2-aminoimidazolium ion contains a highly delocalizable guanidinium system, its pK_a of 8·46 appears to be a little on the low side.

Of the three possible aminothiazoles, only the 2-isomers have been at all extensively studied, largely because they are readily available by ring synthesis (see p. 318). 2-Aminothiazole, with pK_a 5·4, is of the same order of

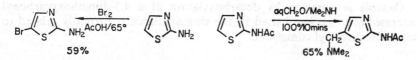

basicity as aniline (pK_a 4·6) and reacts very easily with electrophiles at C 5, as does 5-acetamidothiazole.

Little is known of the amino-oxazoles.

I Synthesis of the 1,3-Azoles

1 RING SYNTHESIS

(a) *From α-halocarbonyl compounds.* As might be anticipated, considerable parallelism emerges from an examination of the major methods available for the synthesis of the oxazole, thiazole, and imidazole ring systems. The most widely used general method employs an α-halocarbonyl compound or its equivalent for C 4 and C 5 and the two heteroatoms come with the other component.

A simple example is provided by the very effective synthesis of 2-aminothiazole, which makes this one of the most readily-available 1,3-

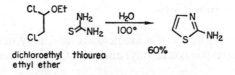

dichloroethyl thiourea
ethyl ether

azole derivatives. Simpler still is the synthesis of 2,4-dimethylthiazole from bromoacetone and thioacetamide.

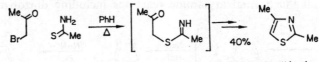

2,4-dimethylthiazole

Thiazole itself is probably best prepared from 2-aminothiazole as follows:

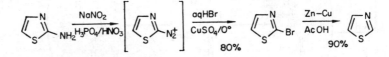

Imidazole can be prepared in good yield directly from chloroacetal, ammonia, and formamide.

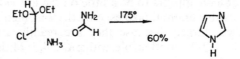

Oxazole is obtained by decarboxylation of a 4,5-dimethoxycarbonyl intermediate: the ketol used in the ring synthesis is formally related to an α-halocarbonyl compound.

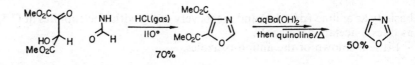

(b) *Other methods.* A widely-used synthesis of imidazoles employs an α-aminocarbonyl compound (or equivalent) in a reaction with thiocyanate anion:

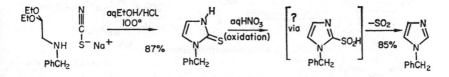

Desulphurization of the intermediate by oxidation is a remarkable process which probably goes *via* a sulphinic acid which then is hydrolysed off as sulphur dioxide.

The classical synthesis of oxazoles centres on the acid-catalysed closure

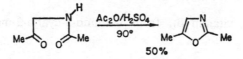

of α-acylaminocarbonyl compounds and formally resembles the furan ring closure (see p. 251).

2   EXAMPLES OF 1,3-AZOLE SYNTHESES

(a) *Histidine*, an essential amino acid.

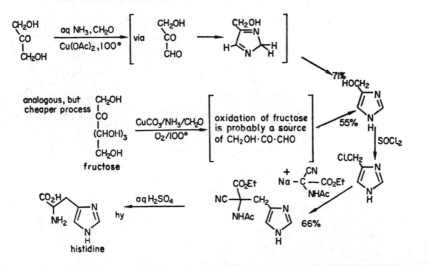

(b) *5-Ethoxy-4-methyloxazole* and its conversion into pyridoxine.

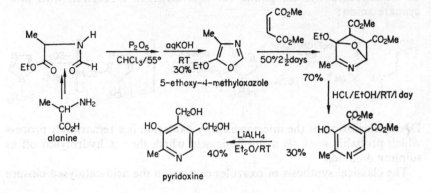

5-ethoxy-4-methyloxazole

pyridoxine

(c) *Thiamin*, vitamin B_1, from 4-amino-5-cyano-2-methylpyrimidine (see p. 142).

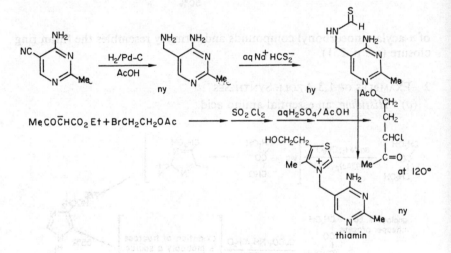

thiamin

27

1,2-Azoles: General Discussion and a Comparison with the Isomerically Related 1,3-Azoles

As in the 1,3-azoles, the lone pair of electrons on the azomethine nitro-

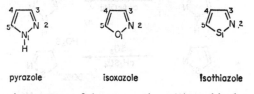

pyrazole   isoxazole   isothiazole

gen of 1,2-azoles is not part of the aromatic sextet and is therefore available for salt formation without disruption of the aromaticity.

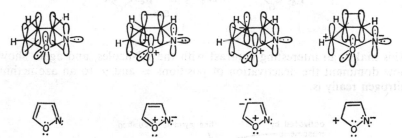

The direct linking of the two heteroatoms has a very marked base-weakening effect, as in hydroxylamine and hydrazine (pK_a: NH₃, 9·2; H₂NNH₂, 7·9; NH₂OH, 5.8): pyrazole with pK_a 2·52 is thus about 4·5 pK_a units weaker than imidazole, and isoxazole with pK_a—2·03 is about 3 pK_a units weaker than oxazole.

Electrophilic Substitution

Pyrazole, like imidazole, has been shown to nitrate by way of the conjugate acid in 95 per cent sulphuric acid, and at about the same rate as the imidazolium cation: this is perhaps surprising, in view of the low basicity of pyrazole.

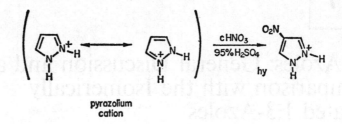

pyrazolium cation

No mechanistic study of isoxazole or isothiazole substitution is yet available.

Electrophilic substitution of 1,2-azoles occurs at C 4, as the selection of examples show:

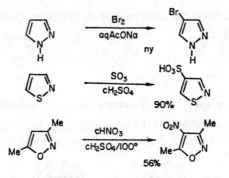

This makes an interesting contrast with the 1,3-azoles, and again shows how dominant the deactivation of positions α- and γ- to an azomethine nitrogen really is.

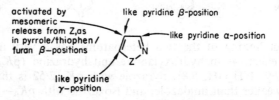

Canonical structures for intermediates of reaction at C 3 and C 5 of course include the very unfavoured positively-charged sextet azomethine nitrogen: canonical structures for the intermediate of reaction at C 4 do not involve the azomethine system at all.

C3 attack
by X⁺

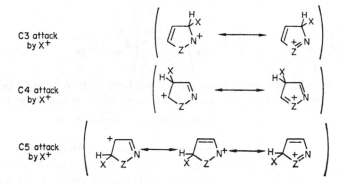

C4 attack
by X⁺

C5 attack
by X⁺

Reactions with Nucleophiles

The 1,2-azoles, like oxazole and N-alkylimidazole, do not react with nucleophiles by substitution of hydrogen. Some reactions involving displacement of halide from C 3 and C 5 in quaternary pyrazolium salts, however, are known.

Ring deprotonation does occur, and reversibly, at C 5 in isothiazoles and pyrazoles. In isoxazoles, C 3 or C 5 deprotonation leads to irreversible ring fission as in the following example:

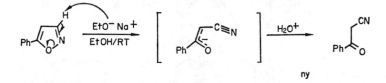

Isoxazolium quaternary salts are particularly susceptible to ring cleavage initiated by deprotonation:

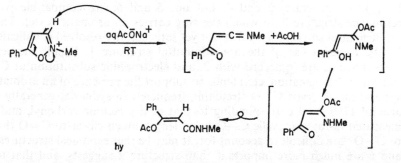

Isothiazole and pyrazole quaternary salts, on the other hand, undergo reversible deprotonation at C 5 under mildly basic conditions, much in the same way as the 1,3-azolium salts do at C 2. Exchange by this process, in isothiazolium and pyrazolium salts, is 10^3–10^4 times slower than for their 1,3-azolium counterparts.

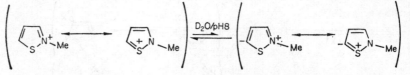

The ylid system derived from the 1,2-azolium salts is seen to be structurally related to that from the 1,3-azolium salts, in the sense that the

3-methylthiazolium ylid:Y=$\overset{+}{N}$Me,Z=CH

2-methylisothiazolium ylid:Y=CH,Z=$\overset{+}{N}$Me

carbon carrying the negative charge is influenced by the positively-charged azomethine nitrogen, either directly (in the 1,3-azoles) or through the intermediacy of a conjugated double bond (in the 1,2-azoles).

Meso-Ionic Systems

There is an increasing number of stable cyclically conjugated heterocyclic systems for which no plausible unpolarized canonical structure can be written: such systems are termed meso-ionic. The most widely studied from among a large number of examples is sydnone, generally formulated as 1 or 2. The more stable of the possible canonical structures are represented

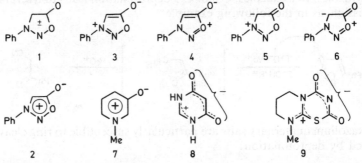

by 3,4,5,6: structures 3 and 4, but not 5 and 6, are compatible with resonance structure 2, in which the ring carries an aromatic sextet. That the structural situation is in fact not yet satisfactorily resolved is indicated by the frequent use of the non-committal structure 1. Dipole moment measurements are in accord with 2, and electrophilic substitution at C 4 (nitration, halogenation, etc.) tends to support the presence of an aromatic system; an observed C—O stretching frequency in sydnones generally of around 1760 cm^{-1} corresponding to that of a γ-lactone carbonyl, and a measured (x-ray) exocyclic C—O bond length much closer to C=O than to C—O$^-$ are difficult to account for. It may be that canonical structures 5 and 6 are much more important than structure 2 suggests, and that the last word has not yet been said.

N-Methyl-3-hydroxypyridine betaine, 7, is likewise a meso-ionic compound, and the simple structure 7 has recently been shown to be inadequate (see pp. 61 and 64); 8 and 9 are further examples of six-membered ring systems which can be termed meso-ionic.

28

Purines: Reactions and Synthesis

In spite of their complexity and of the incomplete state of knowledge concerning their chemical reactivity, purines must be included in any elementary treatment of aromatic heterocyclic chemistry, because of their fundamental importance in life processes.

The treatment of the topic given here separates the chemistry of purine itself and the simple alkyl purines from that of the much better known and biologically important oxy- and amino-purines.

The numbering of the ring system is anomalous and reads as if purine were a pyrimidine derivative.

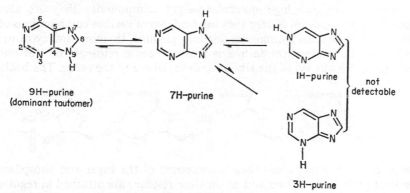

9H—purine
(dominant tautomer)

7H—purine

1H—purine

not detectable

3H—purine

There are in principle four possible tautomers of purine containing an N-hydrogen; in the crystalline state purine exists as the 7H-tautomer, however in solution both 7H- and 9H-tautomers are present in approximately equal proportions.

By far the most important natural occurrence of purines is in the nucleotides and nucleic acids, compounds which perform some of the most crucial functions in fundamental metabolism.

Nucleotides are low-molecular-weight compounds made up of either D-ribose or 2-deoxy-D-ribose, linked by the C 5 oxygen to a phosphate unit, and by the glycosidic C 1 to a nitrogen of one of six heterocyclic systems, of which three are the pyrimidines uracil, cytosine, and thymine (see p. 133), and three are purines adenine, guanine, and hypoxanthine.

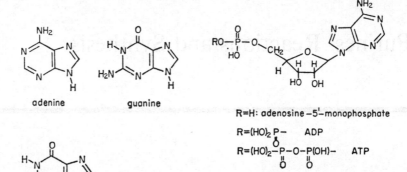

adenine   guanine

R=H: adenosine –5'–monophosphate

R=(HO)₂ P–   ADP

R=(HO)₂–P–O–P(OH)–   ATP

hypoxanthine

Adenosine-5'-monophosphate is a typical nucleotide, which is also known as muscle adenylic acid because it was first isolated from animal skeletal muscle. This nucleotide is related to the key pair adenosine diphosphate (ADP) and adenosine triphosphate (ATP), which play the central role in biological phosphorylation.

Nucleic acids are high-molecular-weight compounds; they are also known as polynucleotides, for they are made up of various combinations of four nucleotide units (containing adenine, guanine, thymine, and cytosine) in which the internucleotide link is that between a phospate unit of one and the C 3 hydroxyl of the ribose or deoxyribose of the other. The back-

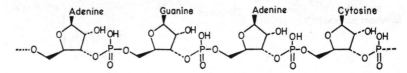

Adenine   Guanine   Adenine   Cytosine

bone of the polymer chain then is composed of the sugar and phosphate units, to which the purine and pyrimidine residues are attached at regular intervals and from which they project. The above formula is part of a ribose nucleic acid (RNA); a chain containing deoxyribose units is called a deoxyribose nucleic acid (DNA).

Nucleic acids occur in every living cell, in which they direct the synthesis of proteins and are responsible for the transfer of genetic information. Quite a lot is already known about their mode of action: what is of particular interest to the organic chemist is that one of the vital aspects of the transfer of genetic information is connected with the reversible longitudinal close association of two nucleic acid strands, which is based entirely on very specific hydrogen bonding between an adenine residue of one strand and a thymine residue in the precisely opposite section of the other strand, and between a cytosine residue on one strand and a guanine residue on the other. This pairing is absolutely specific, for adenine cannot hydrogen-bond with guanine or cytosine, cytosine cannot hydrogen-bond with thymine or adenine, etc.; this specificity is the consequence of the structure, or stereochemistry, or both, of a particular opposing situation, and is illustrated in the following partial structures:

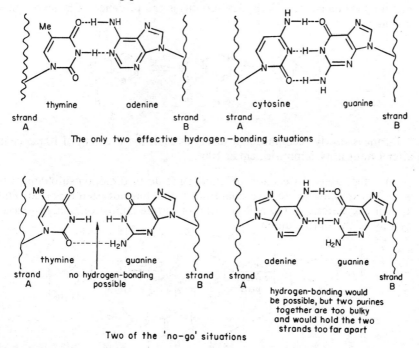

The only two effective hydrogen–bonding situations

Two of the 'no-go' situations

It is amazing to see how heredity and evolution are dependent on two pairs of hydrogen-bonding situations!

The end product of nucleic acid breakdown in birds and reptiles is uric acid (see p. 333). Although in animals this end-product is urea, uric acid is also excreted, albeit in very small quantities. Uric acid must have been one of the very first heterocyclic compounds to be isolated as a pure substance, for it was obtained by Scheele in 1776 from gall-stones.

Mention must also be made of caffeine (p. 331), which is the stimulant present in tea and coffee.

REACTIONS OF PURINE AND ALKYLPURINES

A　Reactions with Electrophilic Reagents

1　ADDITION TO NITROGEN

(a) *Protonation.*　Purine is a weak base (pK_a 2·5). The dominant site of protonation is not known with certainty. Most investigations point to N 1 as the main site for attachment of a proton, but ^{13}C NMR studies suggest that all three possible protonated forms are present in solution and certainly exchange at C 8 (p. 329) confirms the presence of at least some cation 3.

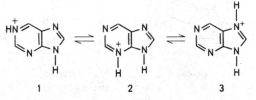

Purine is slowly decomposed in aqueous acid, to the extent of 10 per cent after 1 hour in N sulphuric acid at 100°.

(b) *Alkylation.*　Reaction of methyl iodide or dimethyl sulphate with 9- or 7-methylpurines to give quaternary salts has not been reported. The known alkylations of the N 9-position of purine almost certainly proceed by way of the purine anion.

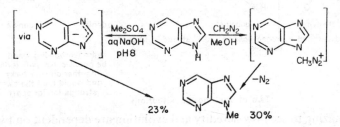

2　SUBSTITUTION AT CARBON

Reactions such as nitration, halogenation, or substitution with any of the milder electrophiles, have not been reported for purine or simple alkyl purines.

B　Reactions with Oxidizing Agents

Reaction with peracetic acid involves attack at N 1.

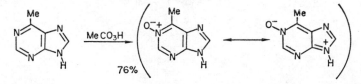

C   Reactions with Nucleophilic Reagents

(*a*)   *Deprotonation.*   Purine, with an acidic pK_a of 8·9, is a somewhat stronger acid than phenol and a much stronger acid than imidazole (pK_a 14.2) or benzimidazole (pK_a 12.3). This relatively high acidity may be mainly the consequence of extensive delocalization of the negative charge in the anion over the four nitrogen atoms. The alkylations of

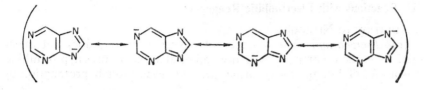

purine described above show that electrophilic addition to this anion occurs at N 9.

Purine is completely deuterated at C 8 in neutral water at 100° in 20 minutes. This process proceeds by the hydroxide-catalysed deprotonation of the 7H-purinium cation 3 to yield a transient ylid of the type thought to be responsible for exchange in the imidazole type system (p. 311).

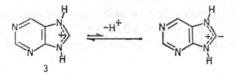

(*b*)   *Attack by hydroxide ion* readily converts 9-methyl purine into 5-amino-4-methylamino pyrimidine. This reaction must proceed by nucleophilic addition to C 8, and has no counterpart in imidazole chemistry.

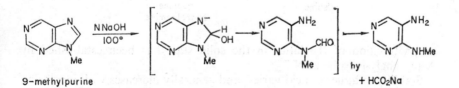

By contrast, purine itself is stable in hot aqueous sodium hydroxide, for, as the purine anion, it is resistant to nucleophilic addition.

D   Reactions with Reducing Agents

Purine is resistant to hydrogenation in the presence of palladium-charcoal, the cation, however, takes up one mole of hydrogen. Purine is also reduced in aqueous sulphuric acid by zinc, though in neither reaction has the product been characterized.

REACTIONS OF OXY- AND AMINO-PURINES

These are tautomeric compounds which exist predominantly as carbonyl and as amino structures, and thus fall in line with the corresponding pyrimidines, imidazoles and pyridines.

A   Reactions with Electrophilic Reagents

1 ADDITION TO NITROGEN

(a) *Protonation*.   The presence of oxygen functions does not affect basicity in these compounds to any appreciable extent, thus hypoxanthine has a pK_a of 2.0, and 8-oxypurine, pK_a 2·6, even though protonation of

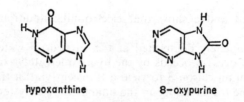

hypoxanthine   8-oxypurine

the latter is likely to be on N 1 or N 3. Aminogroups increase basicity as shown by adenine, with a pK_a of 4·2, and an oxy-group reduces the basicity of an amino-purine, as shown by guanine, with a pK_a of 3.3. The position

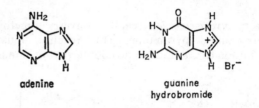

adenine   guanine
                              hydrobromide

of protonation of the latter in the solid state has been established by X-ray analysis to be at N 7.

Stability to aqueous acid varies, and generally increases with increasing substitution. Xanthine is stable to aqueous N sulphuric acid at 100°,

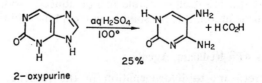

2-oxypurine

whereas 2-oxypurine is completely decomposed in 2 hours with the same reagent.

Amino groups can be replaced by oxygen under conditions of acid-catalysed hydrolysis; some ring-opening occurs as well, this time of the

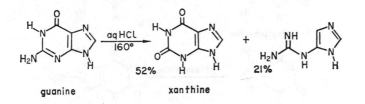

pyrimidine ring. A much better way to convert guanine into xanthine, however, is by the action of nitrous acid.

(b) *Alkylation.* The oxy and amino-purines are usually more easily alkylated than the parent purines. The site of alkylation depends on whether the neutral molecule or the anion is attacked. Thus, with xanthine, the neutral compound is alkylated at N 7 and N 9, whereas the anion

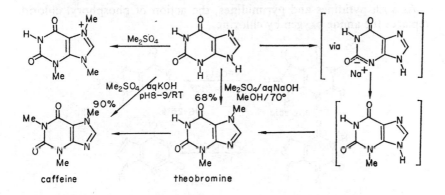

reacts at N 3 first, though this does not constitute a good synthesis of 3-methylxanthine.

Adenine reacts with benzyl bromide in DMF at N 3 only, whereas the conjugate anion reacts with ethyl mesylate to give a mixture of 3-ethyl, 1-ethyl and 9-ethyl adenines. Other alkylating agents, such as $Me_3C.CO.OCH_2Cl$-potassium carbonate attack N 9 exclusively, which is difficult to explain.

2   SUBSTITUTION AT CARBON

Substitution is now possible, though only relatively few examples are known. 9-Methylxanthine is brominated at C 8; theobromine is nitrated also at C 8.

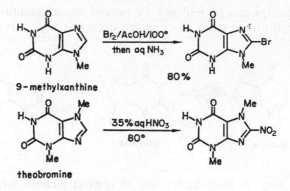

9-methylxanthine

theobromine

B Reactions with Oxidizing Agents

With peracetic acid or pertrifluoracetic acid, N-oxides are obtained; the point of attack varies with structure. Adenine is oxidized at N 1, whereas guanine is oxidized at N 7.

C Reactions of Oxypurines with Phosphoryl Chloride

As with pyridines and pyrimidines, the action of phosphoryl chloride replaces the amide oxygen by chlorine.

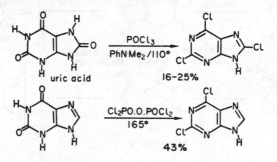

uric acid   16–25%

43%

D Reactions with Nucleophilic Reagents

Deprotonation occurs readily in alkali at ring N-hydrogen, especially when this forms part of an amide grouping. The resulting mesomeric anion, for example, 1, carries the charge mainly on the oxygen.

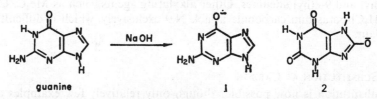

guanine   1   2

Uric acid, the most highly oxygenated purine, is a relatively strong acid, with a pKa of 5·75, deprotonation occurring at N 9, to give 2.

E Reactions with Reducing Agents

Zinc and acid treatment of hypoxanthine results in opening of the pyrimidine ring, presumably by way of reduction at C 2–N 3, and hydrolysis of the aminoacetal function.

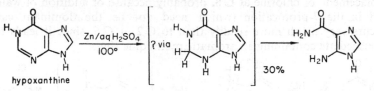

F Reactions with Free Radicals

Recent work has shown that methyl radicals generated by photolysis of t-butyl peracetate effect substitution mainly at C 8, thus adenine gives a 50% yield of 8-methyladenine. This type of reactivity may be involved in certain types of chemical carcinogenesis.

REACTIONS OF HALOPURINES

These, especially the chloropurines, have played an important part in purine chemistry largely because of the ease with which they react selectively with nucleophilic reagents by substitution, and the ease with which they are hydrogenolysed.

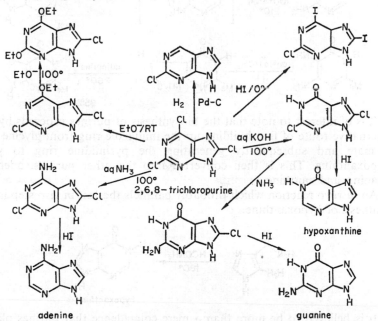

The above conversions of 2,6,8-trichloropurine, easily prepared from uric acid, illustrate the synthetic value of halopurines.

The resistance of chlorine at C 8 to nucleophilic displacement when N 9 is unsubstituted must be the consequence of N 9 deprotonation, the negative charge then shielding C 8 from addition of a nucleophile. Acid catalysed hydrolysis is possible, however, and results in the preferential replacement of chlorine at C 8, probably because of addition of water to C 8 in the 7-protocation (which need not be the dominant cation) occurs very much more readily than to C 2 or C 6, since it leads to an intermediate containing an aromatic pyrimidine ring.

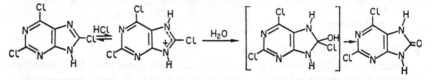

SYNTHESIS OF PURINE COMPOUNDS

1   RING SYNTHESIS

By far the most frequently used approach to this ring system involves first the synthesis of an appropriate 4,5-diaminopyrimidine, followed by formation of the second ring with formic acid or a formic acid derivative.

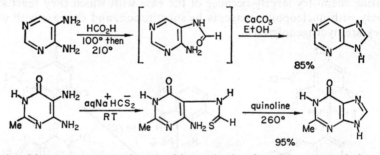

It is of interest to note that the biosynthesis of purines proceeds by the alternative route of first building up the imidazole ring from glycine and formate and subsequently generating the pyrimidine ring to yield hypoxanthine. This is then converted into the other purines, adenine, guanine, and xanthine.

An *in vitro* reaction which probably parallels the last *in vivo* step is the synthesis of hypoxanthine.

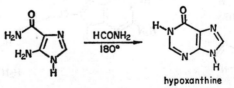

hypoxanthine

It is believed to be more than a mere coincidence that purines play a key role in life processes and that adenine may be synthesized in the laboratory from hydrogen cyanide under mild conditions. In fact adenine can be produced in one step *in vitro* reactions from formamide and phosphorus oxychloride in 36 per cent yield and from hydrogen cyanide and liquid ammonia in 20 per cent yield. These reactions, whilst involving conditions unlikely to have occurred in the prebiotic environment, illustrate clearly the ease with which the purine system can be generated, and allow one to suppose with a reasonable degree of certainty that purines existed before the evolution of life and were incorporated into its mechanism because they were there, and, of course, because they had the right sort of chemical properties.

2   EXAMPLE OF A PURINE SYNTHESIS

Adenosine.

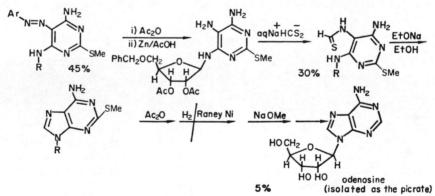

odenosine
(isolated as the picrate)

29

Saturated and Partially Unsaturated Heterocyclic Compounds: Reactions and Synthesis

The main interest in heterocyclic chemistry centres on the heteroaromatic systems, and this book concerns itself mainly with this aspect.

The chemical properties of the remaining heterocycles, a very large number of saturated and partially unsaturated compounds, are so closely similar to the properties of their acyclic or non-heterocyclic analogues that we consider a full treatment unnecessary: in this chapter we shall look briefly at the properties of these compounds, emphasizing in the main those aspects in which they differ from their non-heterocyclic analogues.

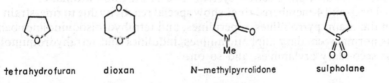

tetrahydrofuran   dioxan   N—methylpyrrolidone   sulpholane

Some of these compounds are very widely used as solvents for organic reactions: tetrahydrofuran (THF) and dioxan for example are often used where ether is not quite satisfactory for reasons of volatility or solvent power. THF and dioxan have to be used with great care because of their relative toxicity and their great tendency to form peroxides when they stand in air; N-methylpyrrolidone and sulpholane are useful dipolar aprotic solvents which fall into the same general category as dimethyl-formamide, dimethyl sulphoxide, and hexamethylphosphoramide (HMPA).

The naming of the saturated and partially unsaturated heterocycles is not completely systematic, in that some have individual names and others are named as derivatives of the aromatic or fully unsaturated equivalents. In the list which follows, the first-given name is that currently used in *Chemical Abstracts*: when there is another name also in use, it is given in second place.

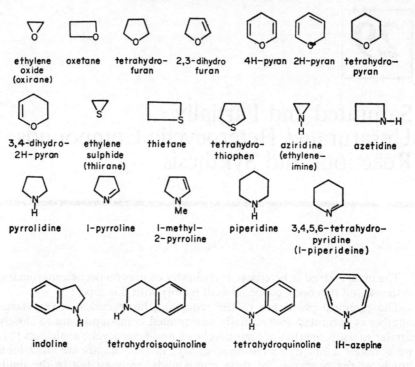

ethylene oxide (oxirane)   oxetane   tetrahydrofuran   2,3-dihydro furan   4H-pyran   2H-pyran   tetrahydropyran

3,4-dihydro-2H-pyran   ethylene sulphide (thiirane)   thietane   tetrahydrothiophen   aziridine (ethyleneimine)   azetidine

pyrrolidine   1-pyrroline   1-methyl-2-pyrroline   piperidine   3,4,5,6-tetrahydropyridine (1-piperideine)

indoline   tetrahydroisoquinoline   tetrahydroquinoline   1H-azepine

Of course this list is not complete, but is just a selection chosen to illustrate the general situation. Note that the seven-membered ring-system azepine is an eight-electron system, and is thus non-aromatic.

The 3- and 4-membered rings show special reactivity due to ring-strain, but in the main, pyrrolidines, piperidines, and tetrahydroisoquinolines behave as normal secondary aliphatic amines, indolines and tetrahydroquinolines as secondary arylamines, and so on.

1   Pyrrolidines, Piperidines, and Derived Systems

The main aspect in which those systems containing nitrogen in a five- or six-membered ring differ from their acyclic counterparts is in the

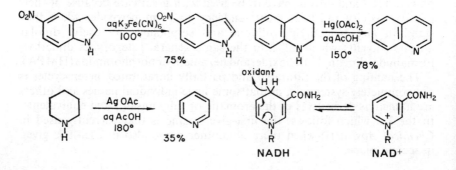

possibility open to them to be easily dehydrogenated to the corresponding heteroaromatic systems, as the examples show.

The dihydroaromatic systems naturally show the greatest tendency to aromatize, indeed one of the important reducing coenzymes, NADH, makes use of this tendency.

Generally speaking, 1-piperideines and 1-pyrrolines exist predominantly in the imine form and not in the tautomeric enamine form; N-alkyl analogues have no alternative but to exist as enamines. They are comparatively resistant to hydrolytic fission of the C:N bond when compared with acyclic imines, but are nonetheless very susceptible to nucleophilic addition to the azomethine carbon. Thus, 1-piperideine and 1-pyrroline themselves exist as trimers formed by nucleophilic addition of

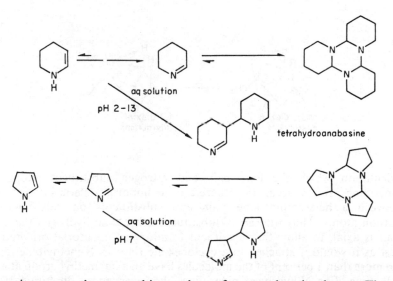

tetrahydroanabasine

one nitrogen to the azomethine carbon of a second molecule, etc. That a small proportion of enamine tautomer is present in equilibrium is shown by the ready self-condensations with C-C bond formation; one unit acts as an imine and the second as an enamine.

The ability of 1-piperideines and 1-pyrrolines to serve as both imine and enamine in such aldol type condensations is the basis of their role in alkaloid biosynthesis. Formed in nature by oxidative deamination and decarboxylation of ornithine and lysine, they become incorporated into alkaloid structures by condensation with other precursor units.

Hygrine is a relatively simple example in which the 1-pyrroline unit, as an imine, has condensed with an acetoacetate unit.

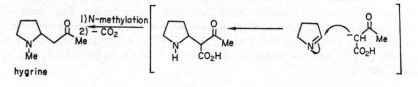

hygrine

Pyrrolidine and piperidine are better nucleophiles than diethylamine, principally because the nitrogen lone pair in each is less hindered. In the heterocycles the two alkyl substituents are constrained back and away from the nitrogen lone pair, and approach by an electrophile is thus rendered easier than in diethylamine where rotation of the C–N and C–C bonds hinders approach.

The pK_a values of pyrrolidine (11.27) and piperidine (11.29) are typical of strong bases; they are slightly stronger than diethylamine (10.98) which may again reflect the less hindered nitrogen atom.

Piperidine and substituted piperidines adopt a conformation analogous to the cyclohexane chair, and in principle, then, there are two possible orientations for the nitrogen atom. Much controversy has centred on the

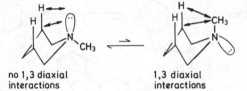

no 1,3 diaxial   1,3 diaxial
interactions   interactions

orientation which the lone pair and nitrogen substituent adopt in piperidines. It now seems that there are no important steric interactions between a heteroatom lone pair and substituents on neighbouring carbon atoms. Thus with an N-substituent other than hydrogen the lone pair is axial, to allow the substituent to adopt an equatorial orientation just as it would if attached to a cyclohexane ring: in N-methylpiperidine no more than 1 per cent of the molecules have an axial methyl group at any one time. For N-hydrogen piperidines, however, more obscure factors must control the orientation of the nitrogen lone pair. Experimental determination of this orientation has therefore entailed the subtle application of a variety of physical methods which have convinced some workers that the lone pair is predominantly equatorial but others that it is mainly axial.

The quaternization of a piperidine in general gives a mixture of products, the proportion of which does *not* depend principally on the preferred conformation of the starting material, but rather on the *relative rates* of axial versus equatorial attack on the nitrogen by the alkyl halide. Axial approach is hindered by the C 3 and C 5 axial hydrogens; on the other hand, equatorial approach forces the existing N-substituent into an unfavourable diaxial interaction with these same substituents. The observed ratio in any particular case depends on an interplay of factors—solvent, structure of the piperidine and of the alkylating agent: methylation of simple piperidines, for example, appears to take place predominantly by axial attack.

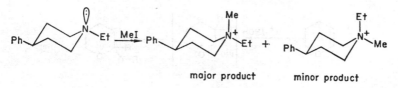

major product   minor product

In the early days of structure determination of alkaloids, in many of which there are saturated nitrogen rings, degradations were used which gave information about the environment of the basic nitrogen. The classical procedure for doing this was the Hofmann 'exhaustive methylation procedure'. This is illustrated as it would be applied to piperidine. What the

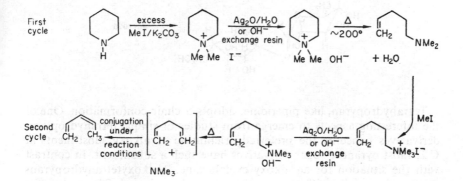

method does is to cleave C–N bonds and eventually remove the nitrogen completely as a volatile tertiary amine. Only one such cycle is necessary to remove a nitrogen which is not part of a ring; a repetition, as in the example above, removes nitrogen when it is part of one ring and a third cycle is necessary if the nitrogen was originally part of two rings. At the end of the process a nitrogen-free fragment is left for study to determine the original carbon skeleton.

In more complex cases, the double bond generated by the ring-opening elimination has very often been ozonolysed to give controlled C–C bond fission and the production of smaller, more easily characterized fragments.

2 The Pyrans and Derived Compounds, and the Reduced Furan Derivatives

4H-pyran has only recently been prepared for the first time: it is extremely reactive, and relatively little is known about it yet. 3,4-Dihydro-2H-pyran and 2,3-dihydrofuran behave as cyclic enol ethers. Dihydropyran is much used in synthesis as a protecting group for alcohols, reaction to an acetal occurring very readily by acid catalysis; the value of such

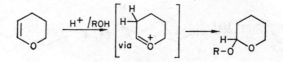

alcohol derivatives lies in that they are very stable to alkaline conditions, and smoothly hydrolysed back to alcohols by dilute aqueous acid.

Quite a lot is known about the hydroxylated tetrahydrofurans and tetrahydropyrans because such ring systems occur in various sugars and sugar-containing compounds; sucrose and RNA (see p. 326) are examples.

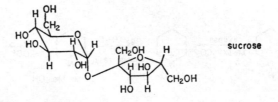

sucrose

Tetrahydropyran, like piperidine, adopts a chair conformation. One of the interesting aspects to emerge from studies of simple tetrahydropyran derivatives concerns the preferred orientation of alkoxy substituents at C 2; most pyranose sugar derivatives have such a substituent. In contrast with the situation for an alkoxy cyclohexane, 2-alkoxytetrahydropyrans exist preferentially with the substituent axially oriented. The reason for

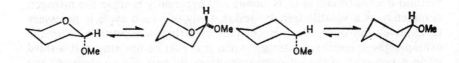

this is that when in the equatorial orientation, there are unfavourable dipole-dipole interactions between lone pairs on the two oxygen atoms, and that the energy-gain, when these are relieved by the C 2-substituent adopting the axial orientation, more than offsets the unfavourable 1,3-diaxial interactions which are introduced at the same time.

Glucose, of which many of the chemical reactions actually involve the small concentration of straight-chain polyhydroxyaldehyde, exists very largely in a cyclic tetrahydropyran form. This illustrates the inherent stability of such chair-like six-membered rings. The propensity for cyclization is a general one for 5-hydroxyaldehydes and ketones and acids; six-membered lactols and lactones can be formed very easily from the straight-chain precursors.

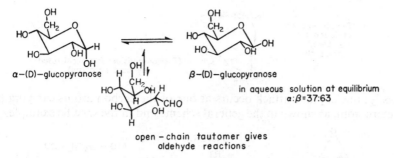

α–(D)–glucopyranose   β–(D)–glucopyranose

in aqueous solution at equilibrium
α:β=37:63

open – chain tautomer gives
aldehyde reactions

Five-membered rings, too, are relatively easy to form; when the glucose
5-hydroxyl is tied up, glucose derivatives can easily be formed in the

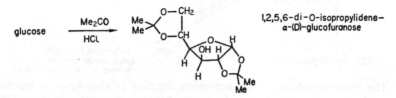

glucose $\xrightarrow[\text{HCL}]{\text{Me}_2\text{CO}}$ 1,2,5,6–di–O–isopropylidene–
α–(D)–glucofuranose

furanose form based on tetrahydrofuran.

3   Three- and Four-Membered Saturated Heterocyclic Compounds

The pK_a of aziridine (7·98) shows it to be an appreciably weaker base
than azetidine (11·29), which in turn is about the same as larger size
nitrogen rings. The low basicity of the small ring is mirrored in the
oxygen series, as measured by the ability of the oxygen heterocycles to
form hydrogen bonds. The explanation for this probably lies in the state
of hybridization which the heteroatom is forced to adopt by the small size
of the ring. The lone pair is in an orbital with less p character than a normal
sp^3 nitrogen or oxygen orbital.

Interestingly, it has been shown that oxetane and thietane are effectively
planar molecules, in contrast with cyclobutane, which is puckered. The

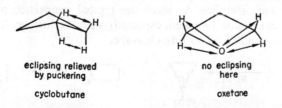

eclipsing relieved   no eclipsing
by puckering   here

cyclobutane   oxetane

removal of some of the eclipsing interactions in the heterocycles is probably
the factor responsible for this difference.

Most of the chemistry of the three- and four-membered saturated heterocyclic compounds is a direct consequence of the strain in such systems. In
addition, there is the ability of the heteroatom to act as a leaving group,
especially with electrophilic catalysis, so they undergo cleavage reactions

Typical reaction
of 3 and 4
membered
heterocycles

(Fission considerably accelerated by acid catalysis)

easily; nucleophilic attack occurs at one of the carbon atoms carrying the heteroatom, as shown in the general scheme and in the specific examples.

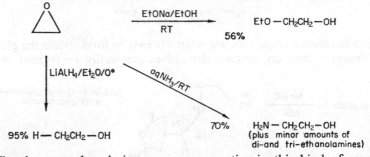

The three-membered rings are more reactive in this kind of reaction than the four-membered rings, because of the greater strain in the smaller systems, thus ethylene oxide reacts with hydroxide ion 10^3 times faster than oxetane does.

Both three- and four-membered hetero-ring cleavages are considerably

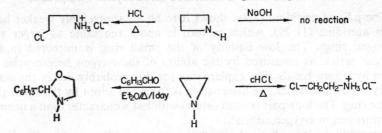

accelerated by acid catalysis and most of the C-nitrogen cleavages in these rings need such catalysis to proceed at all well.

Aziridine and azetidine do have the typical properties of secondary amines, but care must be taken, especially with the former, to avoid conditions which would lead to ring cleavage.

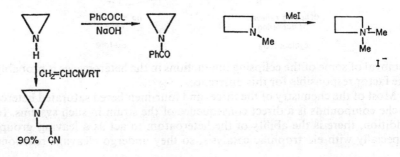

Finally, in this brief survey of the chemistry of the small-ring hetero-cycles, mention must be made of the highly stereospecific elimination pro-cesses which the three-membered ring compounds undergo, in which the

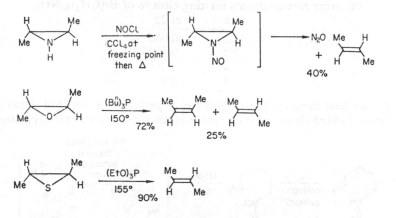

heteroatom is completely removed. Three examples are shown.

4   Ring Syntheses

Many methods have been developed for the synthesis of saturated hetero-cyclic systems, and often these are specific for the heterocycle concerned. One general method, however, is defined by the following formulae:

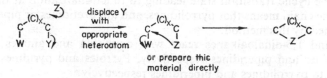

Cyclization takes place by nucleophilic displacement of the functional group at one end of the chain by that at the other end. This process is the reverse of the kind of ring-opening reactions which can be effected easily in the three- and four-membered compounds.

A   SATURATED NITROGEN HETEROCYCLES

Aziridines can be prepared by alkali-catalysed cyclization of 2-halo amines or of a 2-hydroxyamine sulphonate ester.

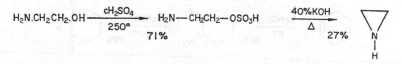

Azetidines can be obtained by an exactly analogous process, but yields are not as good as those for the three-membered rings. Interestingly, the rate of cyclization of halo-amines goes through a minimum at the four-

23*—HC

membered ring size; the five- and six-membered rings are by far the easiest to make.

1st order rate constants for ring closure of $Br(CH_2)_x NH_2$

x	at 25°
2	0·036
3	0·0005
4	~30
5	0·5

A very neat method for the synthesis of pyrrolidines does not require a difunctionalized chain and involves the Hofmann-Löffler-Freytag reaction,

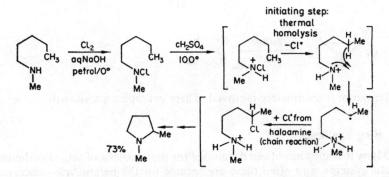

which is a radical process. The example shows how the six-membered size of the cyclic transition state leading to functionalization at the other end of the chain means that pyrrolidines, and not azetidines or piperidines, are formed by this method.

1,4- and 1,5-dihaloalkanes react with ammonia and amines to give pyrrolidines and piperidines respectively. Pyrroles and pyridines can be reduced to pyrrolidines and piperidines respectively.

B SATURATED OXYGEN HETEROCYCLES

The most widely-used method for the preparation of epoxides is unique to them and involves the oxidation of an olefin, usually with peracid.

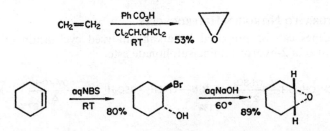

However, a process which is of the general type discussed previously is frequently used. For epoxides this involves the cyclization of 1,2-halohydrins, themselves available from olefins in one step.

Oxetanes can be formed by the base-catalysed cyclization of 1,3-halohydrins, though not as easily as epoxides from 1,2-halohydrins. The

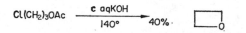

use of the acetate of the alcohol often improves the yield. In contrast to closure of three-membered rings, substitution around the halogen-bearing carbon atom markedly reduces the ease with which cyclization occurs.

Tetrahydrofurans can be obtained by reduction of furans and by cyclization of butan-1,4-diols with acids.

C Saturated Sulphur Heterocycles

Once again the three-membered system can be formed by alkali-catalysed cyclization of a 1,2-difunctionalized precursor, a 1,2-halothiol for episulphides; however, the most widely-used method is unique for episulphides and involves the reaction of an epoxide with thiocyanate ion.

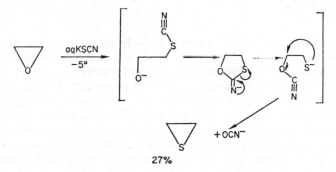

27%

Thietanes, tetrahydrothiophens and tetrahydrothiopyrans can all be prepared in moderate yields by the reaction of the appropriate dihalide with sulphide ion.

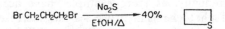

Tetrahydrothiophens cannot easily be prepared by catalytic reduction of thiophens, because desulphurization tends to occur.

30

Photochemistry of Heteroaromatic Compounds

Increasingly over recent years, photochemical reactions of hetero-aromatic molecules, involving either excitation of the heterocyclic nucleus itself or reaction with an electronically excited reactant, have been reported. These photochemical reactions are often of an entirely different type from the ground state reactions discussed in the main part of this book. Electronically excited states have intrinsically different reactivity from ground states because the excitation process leads to a different electron density distribution and a greater conformational mobility due to bond weakening. These factors, together with the high potential energy gained on electronic excitation, allow transformations to products which would be quite unattainable *via* thermal processes. This is why it is appropriate to consider ground state and photochemical processes separately.

Many of these novel heterocyclic transformations seemed, at first, unrelated, but gradually, as more examples have accumulated and as understanding of photochemical processes has increased, patterns have begun to emerge. Generalizations are still more difficult to make than for the main body of classical heterocyclic chemistry, so that the examples given in this chapter are grouped into somewhat arbitrary categories. These have been chosen on the basis of similarities in the overall structural result of the reaction, rather than on a more physico-chemical basis, such as whether a process involves a singlet or triplet excited state or whether the excitation is direct or *via* a sensitizer. For a more mechanistic view-point the student is commended to photochemistry textbooks (see Further Reading).

Since photochemical reactions, other than those where oxygen is directly involved, are always carried out with the rigorous exclusion of oxygen, this detail has been omitted from the experimental conditions given.

A   Substitution Reactions

The group of photochemical reactions which is most similar to some ground state reactions in overall effect are those which involve substitution of ring hydrogen or halogen. Most examples so far are of alkylation or acylation.

(*a*)   *Alkylation* of a pyridine ring can be achieved with an alcohol in the presence of mineral acid: the reaction is of no preparative value, as is evident in the following cases.

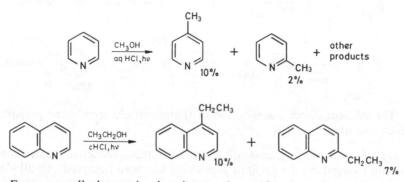

Even a small change in the chromophore of a reactant can however greatly alter the course of such reactions, as is seen in the totally different products obtained under very similar conditions from 2-cyanopyridine.

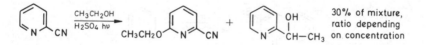

The cyclization of the following indole derivative is one of the very few cases of alkylation of five-membered heteroaromatic rings.

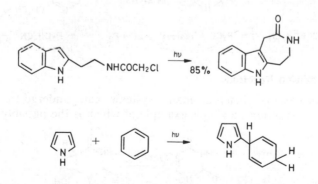

A very interesting and special alkylation of pyrrole is its reaction with photo-excited benzene, which undergoes many analogous additions with a range of non-heterocyclic molecules.

(b) Acylation. Photo-Fries rearrangements in which the acyl group migrates from oxygen or nitrogen to carbon have been reported.

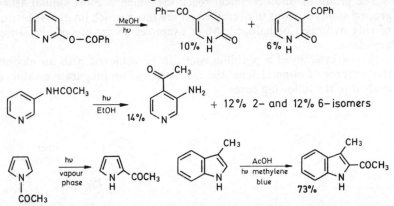

The efficient direct α-acetylation of skatole would seem to be preparatively useful (see p. 262).

(c) Displacement of halogen. In a few instances, reductive replacement of ring halogen from a pyridine β-position has been observed. An alkylative displacement of chlorine in 2-chloropyridine has been shown to involve a one-electron transfer as the initial step in a chain reaction. Halogen at the thiophen 2- or 3-position can be photochemically displaced with attack by carbon of acetone enolate in liquid ammonia solution.

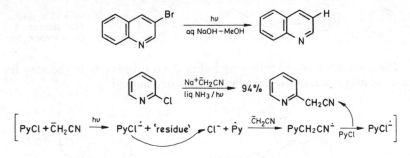

B   Addition Reactions

Photo-excited heteroaromatic systems can undergo a number of addition reactions, a simple example of which is the possibly biologically

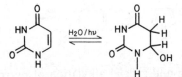

significant reversible 'covalent hydration' of pyrimidines; analogous addition of amines to pyridinium salts is known.

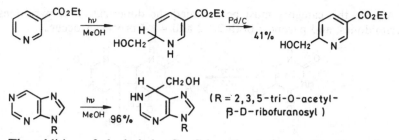

The addition of alcohols by C—C bonding is frequently observed, and is exemplified by the reactions shown above. The alkylation of pyridines described on p. 349 probably proceeds by way of such intermediate adducts.

C Cycloaddition Reactions

There are many examples in which either five-membered or α- or γ-oxo six-membered heteroaromatic rings take part in photocycloadditions. Thus photochemical $2+2$ cycloadditions have been observed for pyridones, pyrones, thiophens, furans, N-acetylimidazoles, and indoles.

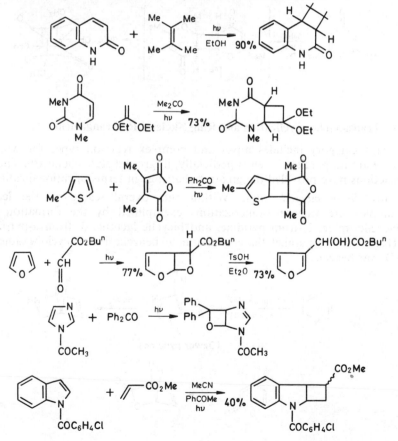

Also in this category must be included the dimerization of pyridones, pyrimidones, and pyrones; both 2+2 and 4+4 processes occur.

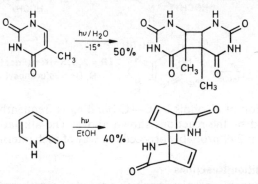

The cage dimer, 1, from 2,6-dimethylpyran-4-one presumably results from a second intramolecular 2+2 addition in an initially formed 2+2 adduct.

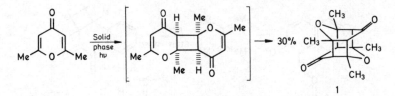

D   Transannular Cyclization and Ring Skeletal Rearrangement

This category includes novel and complex reaction types for which there are no parallels, even superficially, in ground state chemistry. These reactions have provided the greatest challenge in terms of rationalization.

(*a*) *Six-membered rings.* Within this overall category, the least complex are valence isomerizations exemplified by the formation of bicyclic imine, 2, from pyridine, and bicyclic lactone, 3, from α-pyrone. These changes parallel the conversion of benzene into bicyclohexadiene (Dewar benzene).

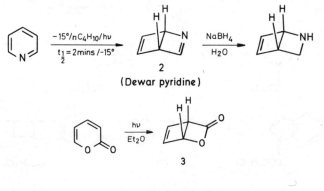

Another type of photocyclisation has been shown to be involved in the irradiation of 3,5-dimethyl-γ-pyrone, which eventually leads to 3,6-dimethyl-α-pyrone.

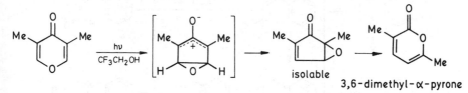

3,6-dimethyl-α-pyrone

Similar [3,1,0] bicyclic intermediates may be involved in the irradiation of charged species such as pyridinium or pyrylium cations, though here the alternative tricyclic oxonia or azonia benzvalene intermediates would also explain the observed products.

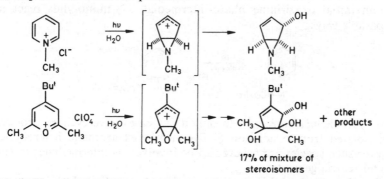

17% of mixture of stereoisomers

Indeed such tricyclic species, e.g. 4, have been shown to be involved in benzene photochemistry and have, for example, been invoked to rationalize the formation of a rearranged skeleton in the aldehyde-ketone, 5, obtained by hydrolytic opening of the pyrylium salt, 6, formed from 7 after irradiation.

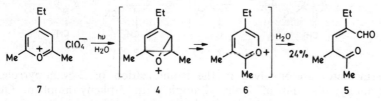

Similar aza-benzvalene structures also allow interpretation of the isomerization of pyrazines to pyrimidines.

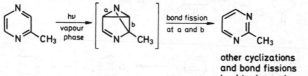

other cyclizations
and bond fissions
lead to 4-, and
5-methyl pyrimidines

Azine N-oxide photochemistry has been more extensively studied than most other aspects of heterocyclic photochemistry. Products, in propor-

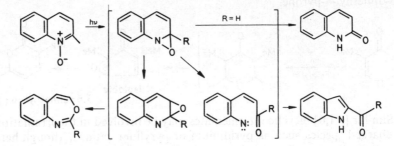

tions depending on solvent and substituents, can be visualized as deriving from an initial oxaziridine photo-intermediate. N-imino-ylids react in comparable ways.

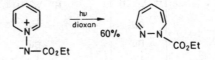

(b) *Five-membered rings.* In only a few cases have cyclized products been isolated from irradiation of five-membered aromatic heterocycles, however such species are believed to be involved as intermediates in ring skeletal rearrangements.

When 2-cyano-1-methylpyrrole is irradiated in methanol, the presumed bicyclic photo-product, 8, is partially trapped by methanol and 9 can be isolated. This observation bears directly on the supposition that analogous

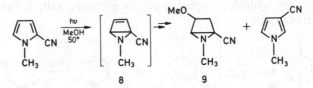

intermediates are involved in the isomerization of 2-cyanopyrrole to 3-cyanopyrrole and of 2-phenylthiophen to 3-phenylthiophen. Other evidence consistent with such an interpretation is the finding that the

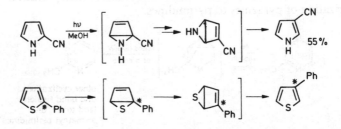

phenyl group remains attached to the same carbon atom in this latter rearrangement. These transformations are visualized as the hetero-atom 'walking' around the four-membered ring in a series of suprafacial 1,3-sigmatropic migrations.

Pyrazoles are transformed into imidazoles by irradiation. Here again a 4,3-fused-ring photo-intermediate, around which the nitrogen 'walks', can explain the observed products.

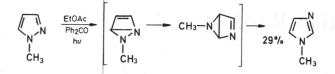

Such rearrangements can even involve annelated benzene rings.

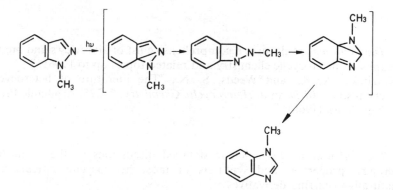

In order to explain the formation of pyrroles during the irradiation of thiophens and furans in the presence of amines, it becomes necessary to postulate, in addition, that the bicyclic photoproducts can be in equi-

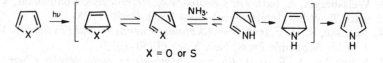

X = O or S

librium with cyclopropenyl aldehydes or thioaldehydes, at which point O or S exchange with nitrogen can occur when reversion to a now nitrogenous bicycle can eventually lead to pyrroles.

Further Reading

The following article gives a comprehensive list of all reviews and monographs on heterocyclic chemistry and related topics up to 1966:
Katritzky, A. R., and Weeds, S. N., 'The literature of heterocyclic chemistry', *Advances in Heterocyclic Chemistry*, **7**, 225, Academic Press, New York (1966).

The following treatises include detailed discussions of the chemistry, physical properties, and synthesis of most heterocyclic systems and naturally-occurring derivatives:
1. Rodd, E. H. (ed), *The Chemistry of Carbon Compounds*, Vols. 4A to 4C, Elsevier, Amsterdam (1957–1973).
2. Elderfield, R. C. (ed.), *Heterocyclic Chemistry*, Vols. 1 to 9, Wiley, New York (1950–1967).
3. Weissberger, A. (ed), *The Chemistry of Heterocyclic Compounds*, Vols. 1 to 29, Wiley (Interscience), New York (1950–1975).
4. Manske, R. H. F. (ed), *The Alkaloids: Chemistry and Physiology*, Vols. 1 to 15, Academic Press, New York (1950–1975).
5. Katritzky, A. R. (ed), *Physical Methods in Heterocyclic Chemistry*, Vols. 1 to 5, Academic Press, New York (1963–1973).

The following reviews and monographs are referred to more than once. The full reference is given here and quoted as the appropriate number in the chapter bibliographies. References to reviews and monographs quoted only once are given in the appropriate chapter bibliography.

6. Marino, G., 'Electrophilic substitution in five-membered rings', *Advances in Heterocyclic Chemistry*, **13**, 235, Academic Press, New York (1971).

7. Norman, R. O. C., and Radda, G. K., 'Free radical substitution of heteroaromatic compounds', *Advances in Heterocyclic Chemistry*, **2**, 131, Academic Press, New York (1963).

8. Rees, C. W., and Smith, C. E., 'Reactions of heterocyclic compounds with carbenes', *Advances in Heterocyclic Chemistry*, **3**, 57, Academic Press, New York (1964).

9. Acheson, R. M., 'Reactions of acetylene carboxylic acids and their esters with nitrogen containing heterocyclic compounds', *Advances in Heterocyclic Chemistry*, **1**, 125, Academic Press, New York (1963).

10. Lyle, R. E., and Anderson, P. S., 'The reduction of nitrogen heterocycles with complex metal hydrides', *Advances in Heterocyclic Chemistry*, **6**, 46, Academic Press, New York (1966).

11. Eisch, J. J., 'Halogenation of heterocyclic compounds', *Advances in Heterocyclic Chemistry*, **7**, 1, Academic Press, New York (1966).

12. Katritzky, A. R., and Lagowski, J. M., *Heterocyclic N-Oxides*, Methuen, London (1971).

13. Ochiai, E., *Aromatic Amine Oxides*, Elsevier, Amsterdam (1967).

14. Den Hertog, H. J., and Van der Plas, H. C., 'Hetarynes', *Advances in Heterocyclic Chemistry*, **4**, 121, Academic Press, New York (1965).

15. Katritzky, A. R., 'Electrophilic substitution of heterocyclic aromatic compounds with six-membered rings', *Angewandte International*, **6**, 608 (1967).

16. Illuminati, G., 'Nucleophilic heteroaromatic substitution', *Advances in Heterocyclic Chemistry*, **3**, 285, Academic Press, New York (1964).

17. Shepherd, R. G., and Frederick, J. L., 'Reactivity of azine, benzoazine and azinoazine derivatives with simple nucleophiles', *Advances in Heterocyclic Chemistry*, **4**, 146, Academic Press, New York (1965).

18. Leffler, M. T., 'Amination of heterocyclic bases by alkali amides', *Organic Reactions*, **1**, 91, Wiley, New York (1942).

19. Schofield, K., *Heteroaromatic Nitrogen Compounds: Pyrrole and Pyridine*, Butterworths, London (1967).

20. Smith, G. F., 'The acid-catalysed polymerisation of pyrroles and indoles', *Advances in Heterocyclic Chemistry*, **2**, 287, Academic Press, New York (1963).

21. Dean, F. M., *Naturally occurring oxygen ring compounds*, Butterworths, London (1963).

The *following chapter bibliographies* make reference to the general reviews, detailed above, by number and also include, in full, references of relevance to the chapter in question.

Chapter 1

Reference: 5
Cook, M. J., Katritzky, A. R., and Linda, P., 'Aromaticity of Heterocycles', *Advances in Heterocyclic Chemistry*, **17**, 256, Academic Press, New York (1974).

358   HETEROCYCLIC CHEMISTRY

Streitweiser, A., *Molecular Orbital Theory for Organic Chemists*, Wiley,
    New York (1961).
Zahradnik R. and Koutecky, J., 'Theoretical studies of physico-chemical
    properties and reactivity of azines', *Advances in Heterocyclic Chemistry*,
    5, 69, Academic Press, New York (1965).
Spiteller, G., 'Mass spectrometry of heterocyclic compounds', *Advances
    in Heterocyclic Chemistry*, **7**, 301, Academic Press, New York (1966).

Chapter 2
References: 1, 2, and 3.

Chapter 3
References: 5, 15, 16, and 17.

Chapter 4
References: 1, 2, 3, 5, 7, and 9–19.
Eisner, U., and Kutham, J., 'The chemistry of dihydropyridines', *Chemical
    Reviews*, **72**, 1 (1972).
Abramovitch, R. A. and Saha, J. G., 'Substitution in the pyridine series:
    effect of substituents', *Advances in Heterocyclic Chemistry*, **6**, 229,
    Academic Press, New York (1966).
Thomas, K. and Jerchel, D., 'Introduction of substituents into the pyridine
    ring', in *Newer Methods of Preparative Organic Chemistry*, W. Foerst
    (ed.), Verlag Chemie, Gmbh., Weinheim/Bergstr. (1961).
Badger, G. M. and Sasse, W. H. F., 'The action of metal catalysts on
    pyridines', *Advances in Heterocyclic Chemistry*, **2**, 179, Academic Press,
    New York (1963).

Chapter 5
References: 5, 15, 16, and 17.

Chapter 6
References: 1–5, 9–13, and 15–18.
Manske, R. H. F., and Kulka, M., 'The Skraup synthesis of quinolines',
    Organic Reactions, **7**, 59, Wiley, New York (1953).
Popp, F. D., 'Reissert compounds', *Advances in Heterocyclic Chemistry*,
    9, 1, Academic Press, New York (1968).

Chapter 7
References: 1–5, 9–13, and 15–18.
Bobbitt, J. M., 'The chemistry of 4-oxy- and 4-keto-1,2,3,4-tetrahydroiso-
    quinolines', *Advances in Heterocyclic Chemistry*, **15**, 99, Academic
    Press, New York (1973).
Popp, F. D., 'Reissert Compounds', *Advances in Heterocyclic Chemistry*,
    9, 1, Academic Press, New York (1968).

Whaley, W. M. and Govindachari, T. R., 'The preparation of 3,4-dihydroisoquinolines and related compounds', *Organic Reactions*, **6,** 74, Wiley, New York (1951).

Whaley, W. M. and Govindachari, T. R., 'The Pictet-Spengler synthesis of tetrahydroisoquinolines and related compounds', *Organic Reactions*, **6,** 151, Wiley, New York (1951).

Gensler, W. J., 'The synthesis of isoquinolines by the Pomeranz-Fritzsch reaction', *Organic Reactions*, **6,** 191, Wiley, New York (1951).

Chapter 8

References: 2 and 3.

Thyagarajan, B. S., 'Aromatic quinolizines', *Advances in Heterocyclic Chemistry*, **5,** 291 (1965).

Chapter 9

References: 5, and 15–17.

Chapter 10

References: 1–3, 5, 10, 12, 13, and 15–17.

Cheeseman, G. W. H., and Werstiuk, E. S. G., 'Recent advances in pyrazine chemistry', *Advances in Heterocyclic Chemistry*, **14,** 99, Academic Press, New York (1972).

Kwiatkowski, J. S., and Pullman, B., 'Tautomerism and electronic structure of biological pyrimidines', *Advances in Heterocyclic Chemistry*, **18,** 338, Academic Press, New York (1975).

Tisler, M., and Stanovnik, B., 'Pyridazines', *Advances in Heterocyclic Chemistry*, **9,** 211, Academic Press, New York (1968).

Chapter 11

References: 5 and 15.

Chapter 12

References: 1–3, 5, 15, and 21.

Balaban, A. T., Schroth, W., and Fischer, G., 'Pyrylium salts. Part I. Synthesis', *Advances in Heterocyclic Chemistry*, **10,** 241, Academic Press, New York (1969).

Mors, W. B., Magalhaes, M. T., and Gottlieb, O. R., 'Naturally-occurring aromatic derivatives of monocyclic α-pyrones', *Fortschritte der Chemie Organischer Naturstoffe*, Ed. L. Zeichmeister, **20,** 131, Springer, Vienna (1962).

Chapter 13

References: 5 and 15.

Chapter 14

References: 1–3, 5, 15, and 21.

Merlini, L., 'Advances in the chemistry of chrom-3-enes', *Advances in Heterocyclic Chemistry*, **18,** 159, Academic Press, New York (1975).

Chapter 15
References: 5, 6 and 19.

Chapter 16
References: 1–3, 5–8, 11, 19, and 20.
Gardini, G. P., 'The oxidation of monocylic pyrroles', *Advances in Heterocyclic Chemistry*, **15**, 67. Academic Press, New York (1973).
Jones, R. A., and Bean, G. P., '*The Chemistry of Pyrroles*', Academic Press, London (1977).
Baltazzi, E., and Krimer, L. I., 'Recent developments in the chemistry of pyrroles', *Chemical Reviews*, **63**, 511 (1963).
Jones, R. A., 'Physico-chemical properties of pyrroles', *Advances in Heterocyclic Chemistry*, **11**, 383, Academic Press, New York (1970).
Lemberg. R., 'Porphyrins in Nature', *Fortschritte der Chemie organischer Naturstoffe*, Ed. L. Zeichmeister, **11**, 299, Springer, Vienna (1954).
Patterson, J. M., 'Recent synthetic methods for pyrroles and pyrrolenines (2H- or 3H-pyrroles)', *Synthesis*, **281** (1976).

Chapter 18
References: 1–3, 5–7.
Steinkopf, W., '*The Chemistry of Thiophen*', Edward Bros., Ann Arbor, Michigan (1944).
Gronowitz, S., 'Recent advances in the chemistry of thiophens', *Advances in Heterocyclic Chemistry*, **1**, 125, Academic Press, New York (1963).
Wolf, D. E., and Folkers, K., 'Preparation of thiopens and tetrahydro-thiophens, *Org. Reactions*, **6**, 410, Wiley, New York (1951).

Chapter 19
References: 1–3, 5, 6, 11, and 21.
Dunlop, A. P., and Peters, F. N., *The Furans*, Rheinhold, New York (1953).
Bosshard, P., and Eugster, C. H., 'The development of the chemistry of furans; 1952–1963', *Advances in Heterocyclic Chemistry*, **7**, 378, Academic Press, New York (1966).
Elming, N., 'Dialkoxy dihydrofurans and diacyloxydihydrofurans as synthetic intermediates', *Advances in Organic Chemistry*, Ed. E. C. Taylor and H. Wynberg, **2**, 67, Wiley Interscience, New York (1960).

Chapter 21
References: 1–9, 11, and 20.
Popp, F. D., 'The chemistry of isatin', *Advances in Heterocyclic Chemistry*, **18**, 2, Academic Press, New York (1975).
Allen, G. R., 'The synthesis of 5-hydroxyindoles by the Nenitzescu synthesis', *Organic Reactions*, **20**, 337, Wiley, New York (1973).
Sundberg, R. J., *The Chemistry of Indoles*, Academic Press, New York (1970).
Heacock, R. A., and Kasparek, S., 'The indole Grignard reagents', *Advances in Heterocyclic Chemistry*, **10**, 43, Academic Press, New York (1969).

Robinson, B., 'The Fischer indole synthesis', *Chemical Reviews*, **63**, 373 (1963) and **69**, 227 (169).

Chapter 23

References: 1–3, 5, and 6.

Cagniant, P., and Cagniant, D., 'Recent advances in the chemistry of benzo[b]furan and its derivatives. Part I: occurrence and synthesis', *Advances in Heterocyclic Chemistry*, **18**, 338, Academic Press, New York (1975).

Iddon, B., and Scrowston, R. M., 'Recent developments in the chemistry of benzo[b]thiophens', *Advances in Heterocyclic Chemistry*, **11**, 177, Academic Press, New York (1970).

Chapter 24

References: 2 and 3.

White, J. D., and Mann, M. E., 'Isoindoles', *Advances in Heterocyclic Chemistry*, **10**, 113, Academic Press, New York (1969).

Chapter 25

References: 5 and 6.

Lakhan, R., and Ternai, B., 'Advances in oxazole chemistry', *Advances in Heterocyclic Chemistry*, **17**, 100, Academic Press, New York (1974).

Chapter 26

References: 1–3, 5, 6, 9, and 10.

Wiley, R. H., England, D. C., and Behr, L. C., 'Preparation of Thiazoles', *Organic Reactions*, **6**, 367 (1951).

Grimmett, M. L., 'Advances in imidazole chemistry', *Advances in Heterocyclic Chemistry*, **12**, 104, Academic Press, New York (1970).

Schofield, K., Grimmett, M. R., and Keene, B. R. T., *Heteroaromatic Nitrogen Compounds: The Azoles*, CUP, London (1976).

Chapter 27

References: 1–3, 5, and 6.

Wooldridge, K. R. H., 'Recent advances in the chemistry of mononuclear isothiazoles', *Advances in Heterocyclic Chemistry*, **14**, 2, Academic Press, New York (1972).

Kochetkov, N. K., and Sokolov, S. D., 'Recent developments in isoxazole chemistry', *Advances in Heterocyclic Chemistry*, **2**, 365, Academic Press, New York (1965).

Slack, R., and Woolridge, K. R. H., 'Isothiazoles', *Advances in Heterocyclic Chemistry*, **4**, 107, Academic Press, New York, (1965).

Kost, A. N., and Grandberg, I. I., 'Progress in pyrazole chemistry', *Advances in Heterocyclic Chemistry*, **6**, 347, Academic Press, New York (1966).

Schofield, K., Grimmett, M. R., and Keene, B. R. T., *Heteroaromatic Nitrogen Compounds: The Azoles*, CUP, London (1976).

Ohta, M., and Kato, H., 'Sydnones and other meso-ionic compounds', in *Non-benzenoid Aromatics*, edited by J. P. Snyder, Academic Press, New York (1969).

Chapter 28

References: 1–3, 5, 16, and 17.

Pullman, B., and Pullman, A., 'Electronic aspects of the chemistry of purines', *Advances in Heterocyclic Chemistry*, **13**, 77, Academic Press, New York (1971).

Lister, J. H., 'Physico-chemical aspects of the chemistry of purines', *Advances in Heterocyclic Chemistry*, **6**, 1, Academic Press, New York (1966).

Chapter 29

References: 1–3.

Lambert, J. B., and Featherman, S. I., 'Conformational analysis of pentamethylene heterocycles', *Chemical Reviews*, **75**, 611 (1975).

Chapter 30

Buchardt, O. (ed), *Photochemistry of Heterocyclic Compounds*, Wiley-Interscience, New York (1976).

Reid, S. T., 'The photochemistry of heterocycles', *Advances in Heterocyclic Chemistry*, **11**, 1, Academic Press, New York (1970).

Cundall, R. B., and Gilbert, A., *Photochemistry*, Nelson, London (1970).

Wayne, R. P., *Photochemistry*, Butterworths, London (1970).

Index

Names, Structures and Numbering of the Simpler Heterocyclic Systems

Pyridine

Quinoline

Isoquinoline

Quinolizinium

Pyrimidine

Pyrazine

Pyrylium

1-Benzopyrylium

γ-Pyrone

α-Pyrone

Chromone

Coumarin

Pyrrole

Indole

2H-Isoindole

Indolizine

Furan

Benzo[b]furan

377

Pyridazine

Thiopyrylium

Thiophen

Purine

Me

Methylthiabenzene

Benzo[b]thiophen

Sym−triazine

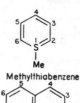

Cinnoline

Quinazoline

Imidazole

Aziridine
(Ethylene imine)

2H−Pyran

Thiazole

Azetidine

3,4−Dihydro
2H−Pyran

Oxazole

Pyrrolidine

Ethylene
Sulphide
(Thiirane)

Pyrazole

Piperidine

Thietane

Isoxazole

Indoline

1,4−Dioxan

Isothiazole

Ethylene
Oxide
(Oxirane)

Morpholine

378

Benzimidazole

Oxetane

Piperazine

1,2,4-Triazole

Tetrahydro
Furan

1,4-Oxazine

Quinoxaline

4H-Pyran

1,4-Thiazine

1H-Azepine

Oxepine

(Hept-)

Thiepine

Azocine

2H-Oxocine

(Oct-)

2H-Thiocine

1H-Azonine

Oxonine

(Non-)

Thionine

(Dec-)

Azecine

Azaundecine

Azadodecine, etc.